In Our Own Image

In Our Own Image

Eugenics and the
Genetic Modification of People

David Galton

LITTLE, BROWN AND COMPANY

A *Little, Brown* Book

First published in Great Britain in 2001
by Little, Brown and Company

Copyright © 2001 by David Galton

The moral right of the author has been asserted.

Permission to reprint extracts from the following poems
has been granted by Faber and Faber Ltd:
Page 78: 'Blessed is the Man' from *Selected Poems* by Marianne Moore.
Page 88: 'Anno Domini' from *Selected Poems* by George Barker.
Page 158: 'Law Like Love' from *Collected Poems* by W. H. Auden.
Page 176: *Nineteen Eighty-Four* by George Orwell (Copyright © George
Orwell, 1949) by permission of Bill Hamilton as the Literary Executor of the
Estate of the Late Sonia Brownell Orwell and Martin Secker & Warburg Ltd.
Extracts from the poem by Emily Dickinson on pages 252 and 255 have been
reprinted by permission of the publishers and the Trustees of Amherst College
from *The Poems of Emily Dickinson*, Thomas H. Johnson, ed., Cambridge, Mass.:
The Belknap Press of Harvard University Press, Copyright © 1951, 1955, 1979 by
the President and Fellows of Harvard College.

A CIP catalogue record for this book
is available from the British Library.

ISBN: 0 316 85592 8

Typeset in Ehrhardt by M Rules
Printed and bound in Great Britain
by Clays Ltd, St Ives plc

Little, Brown and Company (UK)
Brettenham House
Lancaster Place
London WC2E 7EN

www.littlebrown.co.uk

For Merle, Clare, Duncan, Seth,
Julie, Sophie and Jonah

Contents

Part 1
Towards the Genetic Modification of People

Part 2
Which Genetic Markers?

Acknowledgements

Everyone should write their own version of this topic. It is going to affect everyone's life over the next few decades. Were everyone to write a book, each one would turn out differently. The challenge for me was to attempt to see the 'big picture' and draw some firm conclusions from the local patch in which I have been working for the past thirty years (genetic markers and medicine). For this reason I have not included many references within the text. Annotations and footnotes have been avoided so as to keep the prose uncluttered and the intended flow of ideas uninterrupted. In some sections of the text major themes have been borrowed from other people and the list of books and articles in the Sources section gratefully acknowledges this. Much of the newer information has only appeared in articles published in the medical and scientific literature. Readers should contact me if they would like details of these scientific sources.

With regard to Part 1, essentially the background to eugenics, I hasten to add that I am no relation of Francis Galton and am in no way attempting to whitewash the reputation of a distant relative.

The factual content or scientific aspects of the book leave a little room for differing emphasis or interpretation, but more than a third of Part 1 expresses views, opinions and prejudices as to what we should do with the new genetic knowledge, and everyone's perspectives on this will vary. I had better say at the outset where mine come from. A conventional British education from the 1960s onwards led to a science degree in evolutionary biology at University College, London. There Professor J. Z. Young made us write fortnightly essays on such esoteric subjects as 'homology and analogy', 'ontogeny and phylogeny', which to this day I still do not understand properly. I remember him telling me that my essays on these topics were bad, often downright flat; but he expected one to take his criticism in good part. Basic genetics came from J. B. S. Haldane before he and his wife set sail for India. There then followed another science degree in biochemistry, where one highlight was being taught by Francis Crick, of 'double helix' fame. He was visiting lecturer and had already worked out the structure of DNA. As I later found to my cost a major problem in teaching medical students is that of merely keeping them awake. This was not a problem Francis Crick experienced: he was one of the best and most inspiring lecturers I ever heard. He was then working on how to decipher the genetic code – which made for very exciting science.

After my further degree I returned to medicine to complete my training. About half of what we were taught as medical students in the 1960s turned out to be wrong, but the trouble at the time was that our teachers did not know which half it was. So I left medicine to do more science for a doctoral degree at the National Institutes of Health, Washington, DC, where I had the luck to have Bob Scow and Marty Rodbell (later awarded the Nobel Prize for Medicine) as mentors.

The Vietnam War drove many expatriates back to the UK. In my case, I came back to take up medicine at the Hammersmith Hospital, London, and then a final medical-cum-scientific appointment to St Bartholomew's Hospital, London, where I have been ever since.

We are heavily indebted to our teachers and their education helps to make us what we are, but our beliefs and values also come from our experiences in the workplace. The grief of parents losing a young child; the delight of a woman recovering from an illness against all the odds; the courageous resignation of the elderly man with widespread cancer going for weekly radiotherapy; the happiness and pride of a mother bearing her first child; the young lad who will be all right, but gradually and only if he does what we tell him about the daily injections; the old lady of eighty, pallid, confused and bewildered, who neither knows where she is nor how she got there – all these experiences helped to form the opinions and perspectives from which this book has been written. Readers whose major work experience and interests are different – in history, sociology, psychology, commerce, law, ethics, philosophy or theology – may find my text and its contents sadly deficient and in many places downright perverse. I can only apologise to them in advance and invite them to write a version from their own particular viewpoint.

We all have a fatal partiality for our own literary weaknesses; I am therefore particularly grateful to the following friends, colleagues and relations who have pointed many of mine out to me, and helped in many other ways to develop the thoughts and the preparation of this manuscript: Robert Cohen, Mark Caulfield, David A. Galton, Ian Galton, Anthony Galton, John Dickinson, Len Doyal, Paul Rock, John Landon, Brian Shine, Gordon Ferns, John Harris, Ruth Coxon, Clare Galton, Kevin Sommerville, Mike Levene, Andrea Kay, Richard Preece, Katherine O'Donovan, Duncan McNab, Jo Stocks, Melita Thomas, Nick Wald, Kathryn Davies, Richard W. Sharples, Emmanouil Galanakis, John Walker-Smith and Lynne Trenery. David Miller, my literary agent, and Andrew Gordon of Little, Brown were delightfully enthusiastic and interested in this project. I was only sorry that Andrew could not see it through to completion due to his promotion to another publishing house, where I wish him all the success that I know he deserves.

The graphics department of St Bartholomew's Hospital are especially thanked for preparing the line-drawn figures in this book.

Some of Chapters 1, 2, 4, 5, 6 and 7 of Part 1 and Chapters 11 and 12 of Part 2 have been modified from previously published articles by co-authors and myself in the following journals: *Journal of Medical Ethics*, *Quarterly Journal of Medicine*, *Nature Medicine*, *European Journal of Clinical Investigation* and *Reproductive and Genetic Ethics* (see Sources for more details). I am grateful to the publishers of these journals for permission to do this.

Introduction

Good stock to bad is wed
And bad to good, till all the world's cross bred.
No wonder if the country's breed declines,
Mixed metal, Kyrnos, that but dimly shines.

—Theognis, c. 540 BC

'Eugenics' – many people will only have a hazy notion of what the word means, and wonder why anyone should write a book about it. Others will certainly be frightened by the term. The spectre of Hitler's mass extermination camps and, more recently, Serbia's ethnic cleansing programme in Bosnia and Kosovo, rises up under the name of a perverted form of eugenics. The word has acquired a sinister reputation. Recently the editors of an eminently respectable medical journal politely asked me to remove the word 'eugenics' from the title of one of our departmental papers before they would publish it. They were afraid the stigma of the word would attach itself to other researchers in the field.

Call it what you will; but if your aim is to use scientific methods to make the best of the inherited component for the health and wellbeing of the children of the next generation, it is by definition eugenics. Sweeping the word under the carpet or sanitising it with another name merely conceals the appalling abuses that have occurred in the past and may well lull people into a false sense of security. The underlying ideas of eugenics have been with us since the Ancient Greeks. They are persuasive ideas which have made a strong appeal to both right- and left-wing politicians. Brand new and powerful sets of techniques have now been put at the disposal of eugenics.

The word 'eugenics' was initially proposed and defined by a Victorian scientist, Sir Francis Galton (1822–1911). A cousin of Charles Darwin, he was greatly influenced by Darwin's theory of evolution. He defined eugenics to mean 'the science of improving the inherited stock of a population, not only by judicious matings, but by all other means'. The word 'eugenics' comes from the Greek *eu-*, good; *genesis*, creation or birth. It is a sort of polar opposite to the word 'euthanasia' (Greek *eu-*, good; *thanatos*, death). A wider and more up-to-date definition of eugenics might be formulated as 'science applied to the qualitative improvement of the set of human genes transmitted to the next generation'. This definition could be extended to cover methods for regulating population numbers as well as trying to improve the genetic composition of the next generation.

The ideas behind eugenics are not new; as with so many other things the Ancient Greeks were there first. It is impossible to over-state the importance of the impact of Plato and his pupil Aristotle on Western society. It could be said of the European philosophers who came afterwards that they have mainly picked one or two paragraphs from the works of the Greeks and then expanded and analysed them in more detail. The legacy of Plato and Aristotle will continue to be discussed and debated for centuries to come. Both men held decidedly strong views on eugenics, their aim being to provide the city-state with the most able and effective children for the next generation. The methods they proposed are described in Chapter 1. They are bold, radical and, if adopted, would transform family life. They go far beyond the techniques suggested by Francis Galton, who originated the term in the nineteenth century.

Eugenics, then, is a seductive notion and one that has been debated since antiquity. After all, we generally want our children to be healthy and straight-limbed, with clear, enquiring voices.

So how has eugenics acquired its stigma? Eugenics is one of those loaded words – taking in philosophy, ethics, sociology, medicine, genetics, biotechnology – that has acquired many harmful ramifications. Some of the worst of these were hand-picked by right-wing politicians in America, Germany and Scandinavia in the early part of the twentieth century and subsequently misapplied in a grotesque manner.

Initially, laws were introduced to prevent those people considered 'undesirable' from having children. From 1907 onwards, twenty-seven states in the USA passed sterilisation laws to prevent various classes of people from having children. These included the insane, those suffering from epilepsy (the Russian novelist Dostoyevsky would have been included in this category had he lived in the USA) and the feeble-minded (see Chapter 5 [iii]). In some states these laws were applied to habitual criminals and to 'moral perverts'. Most states did not enforce these laws, but in California compulsory sterilisation still managed to reach a total of about 10,000 by 1935. Other countries including Denmark, Switzerland, Germany, Norway and Sweden also passed sterilisation laws for similar categories of people. Between 1935 and 1976 about 60,000 young Swedish women deemed mentally defective or otherwise handicapped were sterilised. Amazingly the sterilisation laws in Sweden remained on their statute books until 1976.

But worse was to follow. When one sees ghastly documentaries about the concentration camps of World War II, the seared and bloated faces of the children, the white skinny trunks of the adults with enormous heads and swollen limbs, one has to ask: 'How did it come to this? How shall we pay for this?' One can at least demand to know whether science was really responsible for this catastrophe. The scientists and geneticists of the Third Reich did indeed justify their extermination programmes on the basis of eugenics. Here is the origin of the stigma. Who would dare to lay on the stones at Belsen or Auschwitz, or on the city gates of Hiroshima, a wreath bearing the inscription *Voluntas Dei*? This phenomenon of man's inhumanity to man is entirely man-made. The science of eugenics (and nuclear fission) has been misapplied in terrible ways, but not primarily by the scientists. Politicians can become totally out of step with the purposes of scientists such as Darwin and Galton. Francis Galton never suggested that state-enforced eugenic techniques should ever be practised. Yet American and European politicians did just that, introducing state-controlled measures such as sterilisation, infanticide and 'euthanasia' (really extermination) for the supposed benefits of society.

So why read a book about eugenics now? The concept has been with us for more than two thousand years, and even the simplest of techniques, such as sterilisation, have been used to commit gross crimes against individual freedoms. This has produced widespread revulsion at the very idea that the state – any state – should interfere with people's reproductive choices, whether this be for the purposes of improving the 'genetic stock' of the next generation or indeed for any other reason. China's current demographic policy of one-child-per-family would fall into this category.

The new genetic technology has produced methods a hundred times more powerful than sterilisation and abortion to regulate which genes are passed on to the next generation. The methods have come as a spin-off from a multinational project to map the position of every gene in the human body and eventually to elucidate the function of each of them (the Human Genome Mapping Project). The term 'genome' here refers to a person's complete set of genes and chromosomes. (See the Glossary for the definition of all such terms.) The project has, and is, providing a wealth of new genetic markers. What these are and how they are used are described in Part 2. Very simply, these markers provide a means of identifying adults, children or embryos who may go on to develop common diseases such as cancer, diabetes or the early dementias. Genetic markers are even being developed to flag up 'personality' traits such as a tendency to aggressive behaviour, to obesity or to depression; and the results of some of these studies have already been published in peer-reviewed scientific journals such as *Nature Genetics*.

How do such techniques affect eugenics? These are fully

discussed in Chapters 3 and 4. Their most dramatic use is in the production of 'designer' babies, where the two techniques of *in vitro* fertilisation and genetic markers are combined. This was the subject of the science fiction film *Gattaca* where 'designer' babies were reared to hold the key positions in society while children born by natural means, termed the 'in-valids', did all the menial jobs befitting a genetic underclass. Subsequent to this film, websites are now available to help you to 'create your own genetically healthy child on-line' – one such may be found on www.genochoice.com. The biotechnology company that has set up this website explores the commercial possibilities of designing your future child. A touchscreen sensor scans your own DNA, reports on your genetic predisposition to a variety of common disorders such as cancer, heart disease and obesity, and then invites you to upgrade your DNA code to create a genetically healthy embryo. Typical fees are $9,000 to correct for sugar diabetes and $11,000 to correct for susceptibility to addiction. Science fiction perhaps – but getting very close to actual practice.

It is natural that parents should want the best for their children. They often make financial sacrifices to provide their child's education, holidays and leisure pursuits; they sacrifice their time and social activities for their children. But it is a lottery as to which of their genes they pass on to their children. This can now be changed. Now they can take steps to prevent transmission of the 'worst' genes that they possess. The 'worst' usually means those genes that can predispose to serious illness. They might even be able to provide their children with some of the 'best' genes that they have; 'best' of course being much more difficult to define.

How this is done and the steps needed to make a 'designer' baby are described in Chapter 3. The techniques are already in use in certain licensed departments in the UK to prevent the transmission of severe inherited diseases such as cystic fibrosis, or a wasting disease of muscle (muscular dystrophy), or a rare fat-storage disease of the nervous system (Tay-Sachs). But it is only one small step for its use to alter the transmission of genes underlying common diseases such as cancer or heart attacks, and another small step to regulate the transmission of genes underlying 'personality' traits such as alcoholism, anxiety neurosis or aggressive social behaviour. Once in the realm of personality traits the vista becomes unending and to many people very frightening. One of the most threatening of these is the new technique of cloning which allows the whole set of genes of a single adult cell to be used to supply the future child. Following on from Dolly, the cloned sheep, in 1997, a private laboratory in the USA proposed to clone human beings for reproductive purposes. The ethics of this has generated much heated debate and discussion, and reproductive cloning has been banned by law in many European countries, in the USA and Japan.

It has been said that the history of society is a record of the struggle of poverty against wealth (de Tocqueville); others have seen it as a struggle of liberty against despotism (Schiller), or of the proletariat against the capitalist class (Marx). But introducing a powerful new eugenic technology affecting future generations that is only accessible to the wealthy or powerful may, in the long run, lead to further social divisiveness and instability.

Governments even now cannot afford to provide freely all the new medical technologies that are of established benefit. Heart transplants, artificial kidney machines, new drugs for cancer all have to be rationed. So the maxims of good government – social justice and equal opportunities for all – may well not apply to the emerging eugenic technologies. Will this lead to a genetic under-class based on poverty? Or alternatively, in view of recent past history, could totalitarian politicians use these techniques to attempt to take control of society? These issues are discussed in Chapter 9.

The penultimate chapter in Part 1 therefore deals with ways and means of regulating the use of these new techniques in a pluralistic society for the benefit of individuals with widely different beliefs and value systems. Good starting points are the statements of the various humanitarian declarations throughout the ages, beginning with one of the earliest. The Oath of Hippocrates (c. 460–377 BC) states amongst other things that 'I will work for the benefit of my patients and abstain from whatever is deleterious or mischievous to them' – a very general statement but a clear indication that the welfare of the particular individual is to be of major concern and not some abstract theory about the 'gene pool', or 'the next generation', or 'the good of society'. Other statements, such as the French Declaration of Human Rights (1789) and the United Nations Bill of Human Rights (1948) are discussed in Chapter 9 in relation to eugenics. Coming to 1997, the Council of Europe produced a Convention on Human Rights and Biomedicine. Some key features they proposed are:

- it is of primary importance to retain the respect and dignity for *Homo sapiens*;
- any genetic procedure requires free and informed consent from the individual concerned;
- everyone has the right to genetic privacy. This means denial of access to information by third parties such as insurance companies, employers, or government departments that could discriminate against a person on genetic grounds.

The Council of Europe also hopes that appropriate sanctions will be put in place to enforce these measures.

These proposals are more than just a set of pious statements. Some people hope that they will eventually be incorporated into domestic law. So far twenty-five countries have signed up to the Convention on Human Rights and Biomedicine. The articles of this convention would be upheld in the European Court of Human Rights in Strasbourg, whose judgements have acquired the force of law in many Western European countries. The Court has become the final place of appeal. Although it has no way of enforcing its decisions directly, it has never been openly defied by any European government and its rulings sometimes prompt changes in domestic legislation. For example, after losing cases before the Court, Britain and France changed their laws on privacy relating to telephone tapping, and Ireland legalised the practice of homosexuality. This is perhaps one path, the legal route, for regulating the eugenics of the future.

We are now at a pivotal time when a new set of eugenic techniques is being introduced (designer babies, gene therapy, gene enhancement, cloning, egg storage and donation, etc.). We have already made a fiasco of our attempts to legislate for the simplest of eugenic techniques – abortion – so how will we handle all the new ones? This book marshals arguments recognising that the uptake and eventual widespread use of the new eugenic technology is inevitable, and argues in favour of the principle that governments and other outside agencies should interfere as little as possible in their adoption. Legislation should mainly be introduced into those situations where calculable harm to others will be the result. Ultimately, the major role for legislation may be to curb the overenthusiastic activities of companies and businesses that promote and sell such eugenic services to the general public.

One last point: the new eugenics has arisen from the merger of two separate disciplines. The first is the new reproductive technology, such as *in vitro* fertilisation, designer babies and cloning. The second discipline is the new knowledge about genes and how they work. Taking the two together has enormously expanded the scope of eugenics. This book therefore falls into two parts. Part 1 deals with the general background to eugenics and then considers some of the scientific issues about genes and how they can be used as markers. Part 2 is more for the reader interested in knowing which genetic markers could be used for the new eugenics.

PART 1

Towards the Genetic Modification of People

Prologue

(I) Flash-Forward to AD 2550

LÖVBORG: . . . this one deals with the future.

TESMAN: With the future? But, good gracious, we don't know anything about that.

LÖVBORG: No. But there are one or two things to be said about it, all the same.

—Henrik Ibsen, *Hedda Gabler*

'Please come in, Ms Karolon-Zeta,' the eugenicist called from the door of his consulting room. 'Do sit down.'

An attractive fair-haired woman entered the room with a quick nervous smile. She was clearly on edge about something; perhaps already expecting some bad news.

'Thank you for the samples of blood from you and your partner last week,' the eugenicist went on. 'We now have a complete gene analysis for you both. I guess you would like to hear the good news first?'

She nodded eagerly. She had waited patiently for seventy years to obtain this permit to have a child. Since the usual life span of adults was now around 500 years there had to be strict controls on birth rates and population numbers. The only other choice was to volunteer to become a colonist for one of the few habitable planets in the galaxy. But since a way of travelling faster than light was never found, the journey to the nearest planet took about 150 years.

The therapist switched on his ocular implant and asked her to activate her own. He then beamed both gene maps on to their implants. Most parts of the maps showed up normally as green. 'Well, there are no serious single gene defects that you need to worry about. Also nothing that needs the use of cloning. Mind you, with our ability to modify and insert new genes as and when we like, this technique is a bit old-fashioned.'

However, a few of the genes on the map showed up in a red script. He now focused on these, and went on: 'The bad news is that any child you might get from the natural route will have a one-in-three chance of developing a learning disability. And if it were to be a boy, he would also have a one-in-five chance of having a mood disorder in later adult life, perhaps even episodes of quite severe depression. I don't know if you are prepared to take these levels of risk.' He was now calling up on the ocular devices a series of three-dimensional models of different children depending on which gene combinations were to be used as blueprints. It was the region of brain connectivity that was being highlighted as abnormal.

The young woman looked hesitantly at the therapist.

'My advice,' he continued, 'would be to go for embryo selection with gene enhancement for one or two of the other weak spots we found in you and your partner's genes. These are mainly in the area of protecting the child from future serious infections such as meningitis and HIV. Also we could do some fine-tuning to correct a tendency to develop early wear-and-tear defects of the hips and knees.'

She half assented, but the jargon was starting to confuse her. 'Can you just explain what that means?'

'Well,' he replied, 'we take a sample of your partner's semen and we would also want a small supply of your eggs from the ovary. We would then select just the right eggs one by one in test tubes and fertilise them with the sperm that have just the right genetic make-up to produce the best children. We save these selected embryos in cryo-refrigerators and destroy the rest. To one of the good embryos, that is the one with the most desirable characteristics, we can add several minor genes to help the baby overcome future infections and the other disabilities. We then implant this embryo into you to start the pregnancy. It would of course be a "natural" embryo but just intensively selected. Oh, and that reminds me, there is one other thing. We spotted another gene sequence in you that won't directly harm the baby but indicates that you will develop eclampsia during the pregnancy.'

'What's that?' Again with a quick nervous smile.

'It means your blood pressure will rise too high and you'll have to rest for about two to three months, preferably in a hospital, to get it treated properly. Let's see now, what is your work?'

'I'm an interplanetary flight controller working on the Biosphere Trans-City.'

'Oh.' He sounded doubtful. 'That's one of the older American space townships, isn't it, put up about 2420?'

'That's right. I help regulate the traffic from Earth to Jupiter's satellites and to the habitable planets beyond. It's mainly for colonisation of that part of the galaxy for people who want to have more than one child and not have to queue for state authorisation to start a family.'

'Well, to be honest, that's not so good,' he replied. 'These older space townships don't have all the health facilities they should. It means you'd have to return to Earth and take an extended period of time off work. Or perhaps you might prefer to develop the baby in an artificial incubator and stay in post. It would save you much of the stress of going through a pregnancy.'

'What about the cost?'

'No problems for you,' he was quick to reassure her. 'Any voluntary action on the part of parents to try for fit healthy children will automatically attract a state birth grant from Oceania. But if you want to leave things to chance and follow the natural route, you'd be well advised to take out a sizeable insurance policy to help cover the costs of all the likely problems that we have discussed.'

She seemed satisfied. 'Thank you. I'll certainly follow your advice and go for embryo selection and incubation methods.'

(II) Flash-back to 360 BC

My Friends, I urge you: get hold of your Greek models, and study them day and night.

– Horace, Epistle 11.3

This conversation, from Plato's *Theaitetos*, takes place in Euclid's house in Megara, about twenty miles west of Athens, in around 450 BC.

SOCRATES: Good heavens, boy, have you never heard that I am the son of a fine buxom midwife called Phainarete?

THEAITETOS: Yes, I have heard as much.

SOCRATES: And have you also heard that I practise the same art?

THEAITETOS: No, never.

SOCRATES: Well I do; but be sure not to tell others . . . You know that midwives by the use of herbal medicines can bring on or relieve the pains of labour just as they wish; or they can render a difficult birth easy, or cause a miscarriage, if desirable, in the early stages of pregnancy.

THEAITETOS: Yes, that is so.

SOCRATES: You may also perhaps have noticed that midwives are expert matchmakers. *They know just exactly which unions between men and women will produce the very best children* [author's italics].

THEAITETOS: Really?

SOCRATES: Oh yes; they pride themselves more on that than on cutting the umbilical cord.

So what has changed in 2,360 years? Instead of selecting partners to produce the 'best' child, we now have the ability to select the egg and sperm to produce the 'best' child – however the word 'best' is defined.

1

At Plato's Academy

Firstly, the legislator must start by endeavouring
to make sure that the infant population shall enjoy the highest
possible state of physical health.

—Aristotle, *Politics*, Book VII

If the grandeur of Rome was its empire, the glory of Greece was its philosophers. Two of the greatest who have ever lived, Plato and Aristotle, worked together as teacher and pupil in the Academy at Athens in 350 BC. It is surprising that Francis Galton, the originator of the term 'eugenics', makes very little reference to either philosopher or to their theories about eugenics. Scientists often try to give themselves more credit for their discoveries and hypotheses than they deserve; they sometimes neglect or try to suppress previous ideas and research on their pet topics. However, the greatest scientists usually acknowledge the work of their predecessors. Isaac Newton, the brilliant mathematician who invented calculus and discovered universal gravitation and the composition of light, wrote to his friend Robert Hooke in 1675, 'If I have seen further it is by standing on the shoulders of Giants before me.'

In one of his lectures Albert Einstein imagined that all the scientists in the world were assembled together for the worship of Grand Unified Theory in the Temple of Science. Then a *deus ex machina* descended in the form of Athena, the goddess of wisdom born from Zeus's brain. First, Athena drove out of the Temple all those seeking personal honour and glory; then she drove out all those doing science from a cold sense of intellectual superiority and the dead vanity of knowing better than the next person; and finally she drove out all those who swore to serve the truth but kept one hand behind their backs for cash or praise. There were very few left by now, but some had a streak of gold in their personality – curiosity, originality and a fascination with what is difficult. They had boldness, a wide-ranging imagination and the courage to think independently. People like that are rare, and Einstein maintained that without them science could never make any of its major advances. They were for the most part odd fellows, solitary, uncommunicative types who had entered the Temple to escape from the hurly-burly of everyday living – to get away from the messy emotional turmoil that we make of our daily lives and personal relationships. They wanted to enter the world of objective reality and pure thought. For them there was neither worldly gain of wealth nor honour; for them it was only a matter of doing the science, and the rest did not concern them.

Francis Galton, who appeared in many ways to be one of the latter types, did not have the excuse of not having access to Greek and Latin literature for failing to acknowledge their contribution. The Greeks and Romans were staple educational fare for the Victorians. He states in a letter that he had read 'Plato's *Republic* and *Laws* for eugenic passages; but they don't amount to much beyond the purification of the city by sending off all the degenerates to form what is termed a colony'. It is not clear why Galton missed many of the eugenic passages and where he obtained this idea of colonisation as a eugenic measure; it does not occur anywhere in Plato's works. In early Greek history (from the eighth to sixth centuries BC) establishing colonies was a way of coping with population expansion and of establishing trading contacts overseas. Perhaps Galton may have been recalling the passage in Book IV of the *Histories* of the Greek writer Herodotus, where Polymnestus, a man of high repute in Thera, had a son named

Battus with a speech impediment (like poor Demosthenes with his mouth full of pebbles). Battus was dispatched by the citizens of Thera with approval from the Delphic oracle to found a colony in Libya. He laid the foundations for the very successful city of Cyrene in eastern Libya. However, Herodotus does not say anything explicitly to suggest that the disability was a motive for exiling him to found the city.

On the other hand, Plato's works do reveal a profound interest in eugenics as a means of providing the city-state with the finest possible children. This was of vital significance for the future of the city to supply men for their army that was continually at war with other city-states. The aim of this chapter is to describe the Ancient Greek views on eugenic methods to improve the quality of their next generation.

Plato's *Republic*

The *Republic* is a young man's book, probably written before 368 BC, when the author was in his forties. According to Plato, to understand society one has to construct it oneself, and his book presents a radical blueprint for the organisation of an ideal city-state. For eugenics the book adopts the policy of ensuring favourable matings. Plato thought it vital for society that correct arrangements should be made for such partnerships.

The first thing Plato would do is to abolish marriage for the elite class. To preserve and multiply the qualities of the elite class, provision should be made for men and women of the highest abilities to form multiple sexual partnerships. He drew analogies with the selective breeding of sporting dogs and horses in order to obtain the best possible stock. In present-day Western society the value of lifelong monogamous marriage as a natural state of affairs is being questioned. Relationships break down, homes have become mobile and collapsible marriages produce a soaring divorce rate. Our evolutionary cousins, the great apes, do not embark on lifelong monogamous partnerships. The obvious alternatives are polygamy, as adopted by Islam; or perhaps it may be safer still to remain celibate. However, Plato thought of a fifth way, an alternative to monogamy, polygamy, celibacy or serial monogamy. Why not institute a series of temporary partnerships for selected men and women of similar natural abilities of intellect, physique and attainments? The aim would be to produce as much diversity of these sought-after qualities as possible from the elite class to pass on to the next generation. Plato never actually defines what these good qualities should be, but leaves it to the reader's imagination.

Plato further suggested that members of the elite classes should only be allowed to breed once they were in their prime; for men, that was upon reaching the age of twenty-five, for women twenty. Inferior members of society should be discouraged from having children. Only the best of the offspring should be kept in the elite class (Plato called them the 'Guardians'), and inferior children should be relegated to the ordinary civilian classes (farmers or craftsmen). These were the general principles.

In practical terms these principles could be implemented by the institution of a marriage festival bringing together suitable young people in the correct age band. Poetry, song and dance would create a romantic atmosphere in which the young couples could 'marry' and cohabit during the period of the festival, which lasted about a month. After this the marriage would be dissolved and the partners would remain celibate until the next festival. The number of marriages at each festival would be at the Ruler's (that is, Plato's) discretion to keep the population numbers constant, taking into account losses caused by war or epidemic disease. Plato feared a decline more than a rise in the birth rate and considered that the civilian classes could breed without restriction so as to keep an average city-state with about 5,000 citizens (as stated in his last book, the *Laws*).

To prevent marriage of the less advantaged members of the elite classes a lottery system would be set up, rigged so that young men who acquitted themselves well in war and other civil duties were given the first opportunities of having a marriage partner for the term of the festival and of producing as many children as possible in subsequent festivals. Inferior youths would draw lots that 'by chance' failed to procure a partner for them. At the end of the festival each marriage would be dissolved but the superior youths would be able to draw by lot another (and different) partner at the next festival. Women would be allowed to bear children between the ages of twenty and forty and men to reproduce between twenty-five and fifty-five, when their physical and mental powers were considered to be at their best. Unofficial unions which produced children would be considered a civil (and divine) offence and appropriate punishments meted out. Men should only have relations with women of a marriageable age if the Rulers had paired them together. Incestuous unions between parents and children were forbidden. But Plato got it wrong with regard to brother/sister unions – there were to be no sanctions against them. Evidently he had not noticed that more deformed children were likely to issue from such unions.

Something like this was tried as recently as the 1940s by the Nazis. One of their lesser-known programmes attempted to 'breed' from an elite Aryan class with blue eyes and blond hair – so as to look as little like Hitler as possible. The Nazis' *Lebensborn* (Spring of Life) policy established Lebensborn homes throughout Germany where blond, blue-eyed SS men and suitable Aryan-looking women were encouraged to breed together. The policy never became firmly established and there were only 700 'employees' in all of the homes. It probably failed to develop as intended because of the overriding demands of the Nazi resettlement and extermination programmes.

Family life in the Greek city-state was to be actively discouraged since it provided a distraction from the business of governing and of defending or extending that city-state. Newborn children were to be taken from their mothers and reared in special nurseries in a separate quarter of the city. Any children born defective would be 'hidden away' in some appropriate manner. This may actually be a euphemism for infanticide. Neither infanticide nor exposure (the abandonment of children in exposed regions, leaving them to the mercy of the elements and wild animals) as practised in Sparta and other Greek cities was openly recommended by Plato for his

Republic. However in Plato's book *Theaitetos*, Socrates, who was the son of a midwife, described a childbirth scene where the midwife suggested exposure of a newborn handicapped baby. The mother resisted, since the child was her first. But it may suggest that the Greeks were keen to accept only 'healthy' children into their society.

Again, abolishing family life has been tried as recently as the 1950s, this time on experimental settlements in Israel – kibbutzim. Children were commonly housed, fed and cared for collectively in a special suite of buildings. Their parents had their own private quarters and could see their own children in the privacy of their rooms in the evenings. Otherwise the children were reared communally. It seemed to work at the time but this system of bringing up children collectively is now marginally practised. An interesting side-effect of this communal childhood was that in later life young adults raised communally considered themselves more like brothers and sisters and tended not to marry amongst themselves but looked outside their kibbutz for partners.

Among the Ancient Greeks and Egyptians, brother/sister marriages were not prohibited. Kings slept with their sisters, mothers with their brothers, and sometimes fathers with daughters. The main purpose of incest between brothers and sisters as practised by the Egyptian royal families was to keep the wealth and the throne within the same family. The Egyptians and the Ancient Greeks probably did not notice the increased frequency of defective children born from marriages of first-degree relatives (that is marriages between siblings, nephews or first cousins). It appears clear from portraits of the Ptolemys, the most notorious example of such incest in Egypt, that some members of the dynasty suffered from inherited disorders, but there is no evidence that the ancients connected such disorders with the practice of brother/sister incest. Children of such consanguineous marriages are much more likely to inherit a variety of defects because both parents possess the same disease-related genes which may well be transmitted in a double dose to their offspring. This is one of the main medical reasons why incest is prohibited in Western societies.

However, if all the newborn children of the elite class in Plato's Republic were brought up communally in a crèche, real brothers and sisters would not perhaps know they were so related, especially as there would be so many different marriage pairs after each festival. Men and women of the elite classes who were past the child-bearing age could form relationships as they wished and these would fall outside the jurisdiction of the Rulers.

Plato's *Laws*

This is a mature work, the last of Plato's dialogues, written in his eighth decade in about 350 BC. It is highly likely that he was at work on the *Laws* for many years during intervals of writing other texts, since it is his longest book and it may well incorporate material from earlier projects. He gives hints of these but seems never to have completed them properly. He has clearly changed his views from the *Republic* and presents a much more practical and extended treatment of the political problems.

Plato's change of heart probably arose from his personal involvement in the practical politics of Syracuse, Sicily, during the period 361–60 BC. He was initially invited there to act as political tutor to the future ruler of Syracuse, Dionysius II, but his efforts went disastrously wrong for many reasons (jealousies, rivalries and power struggles) and he eventually had to escape, not without some personal danger.

In the *Laws* he attempts to frame a model constitution and legislation that might be adopted by a society made up of 'average' Greeks. By now he considers the temporary unions at marriage festivals with a community of wives and children to be unsatisfactory. Instead he would legislate for monogamous marriages, with strict chastity outside marriage. Provision for this should initially be made by arranging sports and dance festivals so the young people of the city could meet each other. If a man of twenty-five or more finds a suitable partner he should submit his request for marriage to the Curator of Laws; if his request is approved, he should in all cases marry at least between the ages of thirty and thirty-five. A girl could marry between the ages of sixteen and twenty. With regard to suitability for marriage a man should select a woman primarily for the good of the city and not just because she takes his fancy. The origins of the bride and the distinction of her family should be carefully scrutinised. Wealth or poverty within the bride's family should play no part in his choice.

Up to the age of thirty-five, any fit young men who refused to marry would be required to pay an annual fine of 100 drachmas if they belonged to the first, wealthiest class; 70 drachmas if to the second; 60 if to the third; and 30 drachmas if to the fourth class. The fines so collected would be dedicated to the upkeep of the Temple of Hera, the goddess of matrimony.

Married couples should make their first concern a matter of eugenics, that is to present the city with the best and finest children possible. They should have antenatal care under the supervision of a Board of Matrons appointed by the Magistrates to superintend the conduct of the married couples. Expectant mothers should assemble daily for a minimum period of twenty minutes at the Temple of Ilithyia, a goddess identified with Artemis or Hera, to make sacrifices and other propitiatory offerings to the goddess. The period of supervision by the Board should be ten years and would provide advice and guidance on all problems connected with childbirth, such as infertility of couples, or advice on contraception if they were producing too many children. Infertile couples could have their marriages dissolved but the relatives of both parties were to have a say in the terms of the separation. Finally, an official register of births and deaths should be kept and be made easily accessible to everyone, all details being clearly recorded on whitened plaster walls. Such a record would be required for the proper observance of the laws fixing the ages for marriage, military service or qualifications for various official posts.

Aristotle's *Politics*

Aristotle, a pupil of Plato, initially studied at the Academy from 367 to 348 BC, but then left to go on his travels. He returned to

Athens in 336 BC and founded the Lyceum. Aristotle is not for those who need no more philosophy than can be printed on the front of a T-shirt. He wrote more than forty-five treatises and books in densely argued prose on subjects ranging from ethics, metaphysics and cosmology to the full *History of Animals*. His intelligence was phenomenal. In the *Politics* he criticises Plato's views on eugenics and then proposes his own. He thinks that the community of wives and children for the elite class is impractical and, if anything, is better suited to the working classes (such as farmers and artisans). If the latter had wives and children in common they would be less closely united by bonds of affection and family ties and would perhaps remain more obedient to the ruling elite. They might also be less rebellious. Also, would mothers in the elite class voluntarily give up their children to be reared in communal nurseries? If some of their inferior children were relegated to the lower classes and eventually discovered their true parents, this might lead to quarrels and recriminations and perhaps incite civil disorder and lead to instability.

Aristotle also criticises Plato's views of leaving the birth rate unrestricted for the lower classes. If this leads to overpopulation, the attendant evils of poverty, crime and revolution would be more likely to follow. In Aristotle's view the birth rate should be regulated even more stringently than it was around 330 BC. He cites in support of this idea Phaedon of Corinth, an early legislator who recommended that the number of land allotments for families and the number of citizens should be kept equal to one another, implying a tight regulation of the birth rate.

Aristotle's own views agree in principle with Plato's that conditions should be organised so as to ensure the highest possible state of health for newborn children. The quality and quantity of the population depend on this and are the first concerns of the city-state. The population is the raw material on which the statesman works. Aristotle would therefore legislate for the following proposals. Strict monogamous marriages should be instituted, with women marrying at about eighteen and men around thirty-seven, the ages at which he considers both sexes to be in their prime.

Pregnant women must take good care of their bodies with regular exercise each day, walking to the Temple of Ilithyia to worship at the altar of the goddess presiding over childbirth. Expectant mothers should be given nourishing food and should remain as tranquil as possible since, according to Aristotle, the embryo derives its basic nature from the mother, rather as plants do from the soil in which they grow.

Laws should be introduced to oppose an unrestricted birth rate, but no children should suffer exposure simply to limit the population. If couples are having too many children abortion must be procured before the embryo has reached the stage of 'sensitive life'. Infanticide should be practised for any children born with deformities.

Other proposed ideal constitutions considered by Aristotle, such as those of Phaleas of Chalcedon or Hippodamus of Miletus, appear to make no provisions for eugenic laws.

Plato's methods of improving the genetic constitution of the ruling elite class are far more original than those of Aristotle and are in accordance with modern genetic theory. A varied choice of partners at serial marriage festivals amongst a selected elite of the population would be expected to lead to an optimum spread of abilities for their children. Like many present-day politicians, Plato practised the systematic deception of the public in the pursuit of his aims. He would rig the lottery system for marriage partners; in Book 3 of the *Republic* he states that deception of the public for the city's welfare is justified for political ends. Plato drew an analogy with Athenian doctors who were prepared to conceal the truth from their patients for their own good.

Aristotle criticised many of the social implications of a community of wives and children, but both he and Plato appear to have been unaware of the possibility of brother/sister (or half-brother/half-sister) matings leading to the appearance of severe inherited disorders in their children. Aristotle mentions without further comment in his *History of Animals* that a stud farm would be of high quality only if the stallions mated with their female offspring. Regarding human beings, he notes that bodily mal-formations in a family, such as cleft lip or palate, may skip one generation only to reappear in the next.

Disappointingly, he never spotted the connection between mar-riages of close relatives and the occurrence of such malformations. Nor did he deduce the existence of any laws of inheritance. However, Plato and Aristotle both recommended some form of legislation to ensure judicious partnerships amongst an elite class, something that would be not be acceptable today. Although arranged marriages are still practised in some societies today, contemporary eugenics has been more concerned with the ques-tionable methods adopted by some politicians of the twentieth century to discourage injudicious mating in the population. These methods include compulsory sterilisation, enforced termination of pregnancy and mass extermination. These are fully discussed in Chapter 5 (iii).

Sparta

Plato's system of eugenics described in the *Republic* has probably never been put to a full and fair test in practice. But something very like it actually existed in Plato's time, in Sparta. Sparta was one of the first states in the world to be run along eugenic lines. It became the most powerful Greek state from about 600 to 350 BC. Society was organised along the lines of Plato's model to select for survival of the fittest men to provide the first-class fighting force which enabled them to dominate Greece for almost 250 years. The supposed legislator for Sparta was Lycurgus, a slightly shadowy historical figure. The following account of how he organised Sparta comes mainly from Plutarch's *Lives*. Plutarch (c. AD 46–119) was Lycurgus's closest contemporary writing about him.

Lycurgus's eugenic measures were directed mainly towards one single aspect, namely that of acquiring military prowess. He prac-tically turned Sparta into an armed camp with a military class ruling with great severity over the subjugated local tribes,

compelling them to work the land for the benefit of their con-
querors. The Spartan elite, men and women, subjected themselves
to the harshest discipline. Gymnastics and war were their major
occupations. Family life scarcely existed and the men lived
together very frugally in a sort of club, always ready for instant
war. Marriage was recognised as merely an instrument for the
production of more soldiers. Their practices of child exposure
and infanticide, aimed ultimately at developing a cohort of good
soldiers, eventually led to a shortage of manpower which proved to
be one of Sparta's major social weaknesses. Laws had eventually to
be passed to encourage population growth by exempting a father
with three sons from military service, and a father of four sons
from all state-imposed obligations, including taxes.

Lycurgus believed that children were not so much the property
of their parents as of the state and therefore legislated that the
'best' men should be the fathers of the next generation. Thus if an
older man had a young wife he could choose a younger man to
father children by her, so hoping to acquire better qualities for his
children. Or if a man admired the children of a married woman he
could, without embarrassment, request her as a bed-mate from her
husband so that he might have children by her too. Newborn children had to be inspected by the elders of the city to see if they had
any defects or deformities – a sort of quality control. If a child had
defects it was killed. If they found the child puny or poorly
shaped, that child was exposed for the night in a chasm near
Taygetus. It was not considered in the public interest to rear a
child who from the outset appeared to be unhealthy. The Romans
also considered malformed babies as ominous; even in modern
times malformations such as hare lips have been viewed with awe
and superstition.

If the Spartan boy passed the inspection he was raised, up to the
age of seven, by his parents and then gradually weaned from
family life. On entering communal life, the child trained for survival in a military garrison. Everything was organised with a view
to military service. Lycurgus outlawed all that he considered needless and superfluous such as the practice of drama, poetry,

sculpture and philosophy. Major 'educational' activities were to be wrestling, running, throwing the discus and javelin, and hunting. Young girls also participated in these athletic exercises to make them strong, healthy and better able to cope with multiple pregnancies.

As the boys grew older they were gradually introduced to military camp life. They were ill clothed, made to sleep on the bare ground, and underwent ritual fights and military excursions to increase their virility and combativeness. Their only personal possessions were a shield and a javelin. To show their courage and bravery they would undergo barbarous initiation rites, and Plutarch claimed that he saw several Spartan youths being whipped to death without complaint as a demonstration of their stoical powers of endurance.

This logical but abhorrent system of breeding and education, more in the nature of animal husbandry, enabled Sparta to remain for a long time the most powerful military city in the entire Greek world and to triumph over many of its rivals, including Arcadia, Argos and Messenia. Eventually Sparta conquered Athens itself after the long struggle of the Peloponnesian War (431–404 BC). For the first time in history a deliberate attempt to build a totalitarian state for a ruling military elite was successful and that state lasted for about 250 years. Fortunately Sparta eventually went into decline and was in turn conquered by the Romans in 146 BC.

A visit to Sparta today shows that little remains from antiquity: no temples, no statues, no sanctuaries. There is an acropolis with a ruined theatre beside it. The Spartans were too preoccupied with military life to leave behind grand or impressive public buildings. Sparta is now primarily remembered for its victory over the Athenian military machine during the Peloponnesian War, as recounted by Thucydides. Athens is a different matter. Its contribution to civilisation – from the philosophy of Socrates, Plato and Aristotle; the drama of Aeschylus, Sophocles, Aristophanes and Euripides; the sculpture of Pheidias, Myron and Praxiteles; the historical woks of Thucydides and Herodotus (who worked in an Athenian colony); to the architecture of Callicrates and Pheidias and the political science of Pericles and Demosthenes – is immeasurable. Comparing the fates of Athens and Sparta may suggest that eugenic legislation for a particular human characteristic such as the 'strongest' or 'ablest' may not in the long run be a very rewarding or civilising approach.

2

The Genetic Revolution: Naming the Parts

To see a World in a Grain of Sand
And a Heaven in a Wild Flower
Hold Infinity in the palm of your Hand
And Eternity in an Hour.

—William Blake, 'Auguries of Innocence'

Genes and genetic markers

William Blake tells us 'to see a world in a grain of sand' but scientists have found a biological world in a structure some thirty times smaller than a grain of sand. They have found it in the nucleated cell.

Everyone begins life as a single cell, a fertilised egg. All the instructions on how to develop into a human being rather than, say, an elephant come in twenty-three packages of genes from the mother and twenty-three packages from the father. These packages are called chromosomes and you have to be given exactly forty-six of them. Genes occur along the chromosomes and are stretches of DNA of varying lengths depending on the size of the protein they make. DNA is shorthand for the molecule *D*eoxyribo*N*ucleic *A*cid. It is a very long thread-like structure arranged in pairs to form a helix (the famous double helix). It is made up of a series of four building blocks called *A*denine, *T*hymine, *C*ytosine and *G*uanine. These chemicals, often called bases, are denoted by the letters A, T, C and G, standing for the

first letter of each of their names (see Figure 1). The genetic part
of the chromosome is the sequence of these bases attached to the
DNA thread. They act as the code to instruct the cell how to make
a protein. Proteins are strings of yet another type of building
block, the amino acids. The importance of the final products of
gene action, the proteins, can be appreciated if you think of a cell
as a factory that imports raw matcrials and thcn manufacturcs
other things for export. Then the major business machinery and
tools of the factory are all made out of proteins. There are many
different types of proteins: some can function as catalysts, called
enzymes, which organise all the chemical reactions that go on in
the cell. Some form the basic structure for the walls of the cell,
and the various membranes within the cell. Others can line the
pores of cell membranes to control what goes through them, acting
as a regulated gateway. Still others act as chemical messengers
(the hormones) to control the activities of other cells. And proteins
can have a host of other functions.

After fertilisation, the egg cell starts to divide into two, then
four, then sixteen, to form a small ball of cells resembling those
simple plankton floating in the expanse of the oceans (Figure 2).
As the cells continue to multiply they start to form the bodily
organs, recapitulating in some mysterious way the course of evolution from single-celled organisms right up to the higher apes.
The simple ball of cells soon resembles a fish with a backbone, a
beating heart and a set of gill slits. The latter are not used for
breathing, as with fish, but turn into structures found in the throat
and special glands in the neck. After further transformations the
embryo starts to look like one of the mammals, with four limbs and
a tail. All the bodily organs of the young human (liver, kidneys,
spleen, lungs, brain, etc.) are eventually laid down to enable independent life from the mother after the birth.

The chromosomes (see Figure 3) are like a collection of recipes
or instructions combined in a unique way for each embryo, carrying half of the genes from each parent from generation to
generation. They carry instructions to run all the working
machinery of the human body: the heart for pumping blood, the

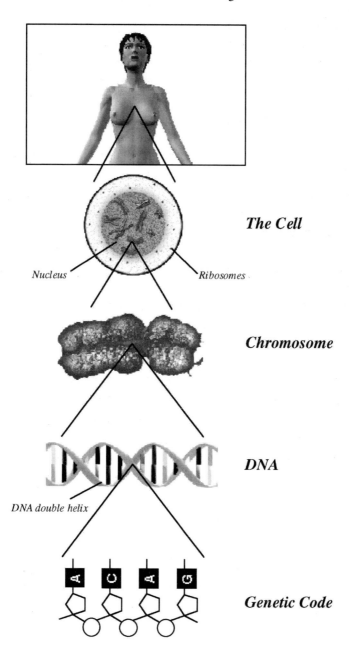

Figure 1. An exploded diagram from the body down to the genetic code.

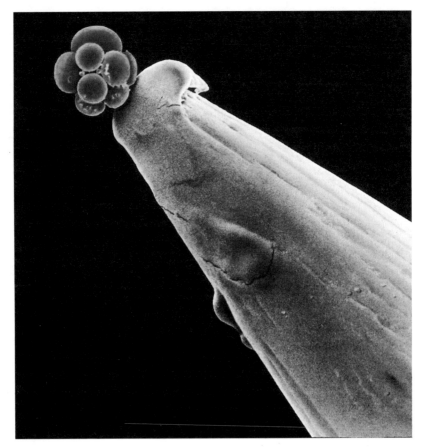

Figure 2. A human embryo at the early cell stage, on the tip of a pin. This ball of cells has the potential to become a person. Should it be given the status of a person? (Science Photo Library)

kidneys for filtering the blood, the lungs for breathing, the skeletal muscles for movement and of course the master-controlling organ, the brain. A photomicrograph of a spread of the forty-six human chromosomes is shown in Figure 3. Each chromosome has the neatness and tightness of a miniature reef knot. The complete instructions (that is the DNA sequence) of chromosome 22 were the first to be announced in December 1999. It is the second smallest chromosome in the body and contains at least 280 genes, although there may be as many as 500. This is the first step to a

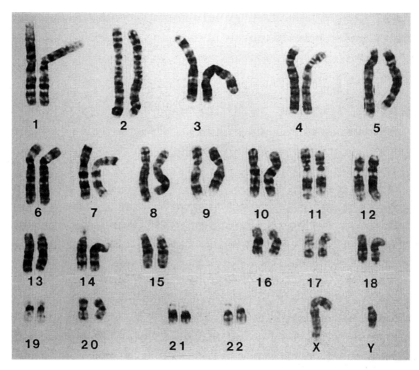

Figure 3. A picture of the twenty-three pairs of chromosomes that make up the human genome. The sex chromosomes on the far right designated X and Y are very different in size; the other pairs are more similar. (Science Photo Library)

complete and accurate working draft of the genes on all the other chromosomes.

Genes are like beads strung out on a long coiled thread of DNA except that there are rather a lot of them – about 30–40,000 genes in each nucleus of each human cell. Most of them are not working at any one time in any one cell. Only about 10–15,000 may be active in making the characteristic proteins for each cell type. To give some idea of the size of genes, if one were to lay all the human chromosomes end to end and scale them to the height of Mount Everest, one gene would be about the length of an ant. The string on which the genes are threaded is also very long. If one were to unravel all the chromosomal threads in the human body and stretch them end to end, the resulting strand would go from the

Earth to the Sun and back again nearly twenty times.* However
the threads on which the genes are strung are tightly packed and
coiled on to spindles, the chromosomes, so as to take up as small a
space as possible. Surprisingly the genes by themselves only
occupy a very small part of the DNA thread, just about 3 per cent
of the total DNA. The rest has been called 'junk' DNA, but this is
almost certainly a misnomer. 'Junk' DNA is probably intimately
involved in packaging and duplicating the genes as well as controlling their activity.

Apart from their chemistry, genes unfortunately have also come
to have a political dimension. Because there are inherited differences in such a highly visible feature as skin colour between
sub-Saharan Africans and north Europeans, wild assertions have
been made that other bodily organs must be genetically different
too. Muscles – Africans run faster; brains – Africans are less intelligent; ears – Africans are better at rhythmic music; reproductive
organs – Africans are better at sex, etc. The fallacy of the reasoning is all too obvious on all accounts (logical, scientific and
humanitarian), but these tired old clichés pursue a life of their own
and have led to innumerable political disputes.

The scale and complexity of the genes do not lend themselves
easily to over simplification, which much of this chapter must
necessarily be. Superficiality means that we have to hurry on
giving half-answers to important issues. It can be worse than
ignorance if one does not realise that one is just skimming the
surface of things. To misunderstand the subject matter is to
admit that one has not looked deeply enough. If you want to dig
deeper, good books for this are listed in the Sources section at the
end of the book. On the other hand, if you are a non-scientist and
dislike this stuff or find it slightly boring, perhaps you can count
yourself lucky. One cannot read at a fever pitch of excitement all

---

* Each cell has a total length of about two metres of double helix in the nucleus.
From the number of cells in the body this makes approximately 6×10^{12} metres
of DNA. The distance from Earth to the Sun is about 1.5×10^{11} metres.
Therefore the double helix would stretch there and back about twenty times!

the time; the boring patches allow you to take a rest before the next section.

Cracking the code of life

The amount of recent interest in cracking artificial codes is surprising. The code breakers of World War II have generated enormous public interest in the form of books (*Cryptonomicon*), films (*Enigma*) and various television series. The recent exhibition at the British Museum in London, 'Cracking Codes', also drew a large audience. Yet few people have any inkling how the code of life was cracked. It turned out to be as difficult and as interesting as deciphering the Rosetta stone.

In 1799, during the Napoleonic occupation of Egypt, a curious black basaltic stone about three feet long was unearthed at the town of Rashid (or Rosetta), some thirty miles from Alexandria. It now resides in the British Museum, having been appropriated after the British drove the French out of Egypt.

The stone bears three sets of inscriptions praising the benefits conferred by Ptolemy V in 205–180 BC. The three texts of the stone from top to bottom are: Egyptian hieroglyphic, a cursive form of hieroglyphic called demotic, and Greek. The problem was to decipher the meaning of the Egyptian hieroglyphics from the other two texts, especially the Greek. Unlike the code of life, which is internally consistent within each of its two 'alphabets', it turned out that Egyptian hieroglyphics can act as alphabetic, phonetic, logographic (signs standing for full words) or even pictographic signs standing for a whole idea or object. This made it an exceedingly difficult problem to solve but, eventually, thanks to the work of Young and Champollion, most of the hieroglyphic signs were assigned to their Greek equivalents. They confirmed that the translation went from the Greek script to the Egyptian hieroglyphs and not the other way round as had been previously thought.

When it comes to the code of life, the problem was to decipher how the four-letter alphabet of the DNA (A, T, C, G) is translated into the exact structure of a particular protein. Proteins are made up of twenty building blocks (the amino acids) and the problem resolves into how the genetic code specifies the exact linear order of these building blocks. The code of life is contained in three different sets of molecules (DNA, RNA and transfer RNA) and uses two different alphabets. Working this out provided the key to understanding how the sequence of bases in the DNA determines the sequence of the building blocks in proteins. Comparing the 'sentence structure' of genes to the grammatical syntax of the Rosetta stone is more than just a simile. It is in fact a very precise and accurate metaphor. The Rosetta stone is a stretch of digital information, written in a linear, three-dimensional script and defined by a code that translates a small alphabet of signs into a large dictionary of meanings determined by the order of the symbols. So is the genetic code. Physicists such as Galileo and Newton thought of the language of the universe in terms of mathematics because so many of the natural laws can be expressed by mathematical formulae. For some people these simple and elegant mathematical laws of Nature comprise the best evidence for a superior purpose working throughout the universe. They appear to defy any other rational explanation – why use just this particular set of laws for constructing the universe? After looking at Figure 4 others might say that God is a better grammarian than mathematician. Still others think that this sort of speculation is a product of approaching senility.

The three key molecules in the code of life are very similar: the DNA containing the genetic code is written from a four-letter alphabet. The RNA (shorthand for *Ribo–Nucleic Acid*) acts as a messenger between DNA and proteins. It is very similar to DNA, containing a four-letter alphabet with just one letter different from the DNA code. The code of RNA acts as an intermediary or messenger in the translation of the DNA into the ordering of the twenty amino acids of proteins to give them their characteristic properties. The third alphabetic molecule, the transfer RNA, is required to align each of the amino acids in the correct order, as specified by the messenger RNA. Complexity is not a crime but it

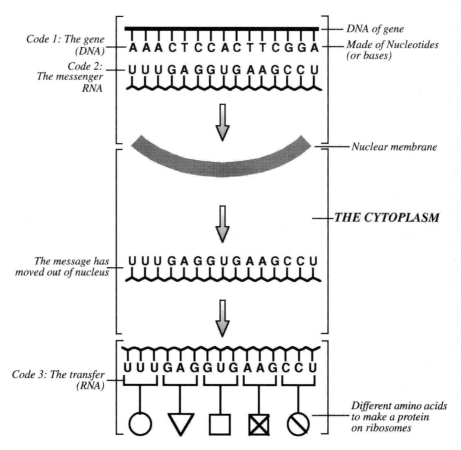

THE THREE CODES   *CELL COMPARTMENTS*

THE NUCLEUS

Code 1: The gene (DNA) — **A A A C T C C A C T T C G G A**

Code 2: The messenger RNA — **U U U G A G G U G A A G C C U**

DNA of gene
Made of Nucleotides (or bases)

Nuclear membrane

THE CYTOPLASM

The message has moved out of nucleus — **U U U G A G G U G A A G C C U**

Code 3: The transfer (RNA) — **U U U G A G G U G A A G C C U**

Different amino acids to make a protein on ribosomes

Figure. 4. The Alphabet of Life: how genes code for proteins.

can be so badly presented as to become lost in murkiness. I hope the following paragraphs, combined with Figure 4, are free from the latter.

Usually one gene provides the code for the manufacture of one particular protein, and deciphering the code for this has been one of the most exciting and fascinating chapters in the development of modern genetics. It is obvious that one of each of the four DNA bases could not code for one amino acid, because this would only allow proteins to be made with four amino acids. We know that proteins comprise twenty amino acids. If you take combinations of two bases at a time this still only gives sixteen possible codes (CT, CA, CG, TA, TG, etc.) and this is not enough to code for the twenty amino acids. It would leave four amino acids with no code. It actually turned out to be a three-letter code (CCA, CCG, AAG, GGA, etc.) that gives sixty-four possible combinations of such three-letter 'words' (that is 4×4×4 = 64). This is more than enough for the number of amino acids and leaves some triplet codes remaining for punctuation signals. So the bases A, T, C and G, corresponding to the four bases of the chemical building blocks of DNA, when taken three at a time, were found in practice to code for each of the twenty amino acids.

The amount of information actually coded in the genes is phenomenal. The four-letter alphabet of DNA can give three billion combinations of 'words' that make more than 30,000 different genes of varying lengths, from less than twenty up to thousands of bases. Less than six-millionths of a gram of DNA contain more information than ten volumes of the *Complete Oxford English Dictionary*, weighing more than ten kilograms. If you assemble all the three-letter words of the genome it comes out to be longer than 800 copies of the Bible; and there are billions of copies of this in each one of us. It is an awe-inspiring perspective on the molecular complexity of our life.

How does the code actually work? Look at Figure 4 on the previous page and I shall try to read it through for you. In the figure, the solid top black line represents the backbone of the thread–like DNA molecule in the nucleus and shows a small part of a gene. Jutting down from the backbone are a series of letters: A, A, A, C, T, etc. These are the bases of DNA arranged in an exact order of letters and are part of the genetic code, labelled code 1, on the left-hand side of the diagram. The bases function as a surface mould or template for another set of molecules, the RNA bases, to line up and then bond together to form the messenger RNA (code 2). This is the dark wavy line in the figure with projection upwards of bases, U, U, U, G, A, etc., that exactly match with the bases of DNA – a phenomenon called complementary base-pairing. This means that only certain bases can pair up with each other depending on the laws of chemistry.

The messenger RNA then falls off the mould of the gene and travels out of the nucleus in the direction of the arrows to the cell-sap, called the cytoplasm. Notice that the order of the bases in the messenger RNA is exactly specified by the order of bases in the DNA. The messenger RNA then joins up with particles in the cytoplasm called ribosomes. These cellular particles are shown in Figure 1 (page 23). Code 3 is found on the ribosomes and is a set of adaptor or transfer molecules made out of the same bases as RNA (the letters are not included in the diagram because of lack of space). These transfer molecules are made of just three bases and they line up and bind strongly to their obligatory partners on the messenger RNA. Now each transfer molecule carries one amino acid, so this means that the amino acids are assembled in an order initially specified by the line-up of bases on the DNA of the gene. The amino acids can then bond together in the right sequence to form a protein. This is how the DNA code is translated into a protein product (the symbols ◯ ▽ ☐ ⊠ ⊘ in Figure 4 represent the different amino acids).

Before you can fully understand genes and how they work you have to learn a whole new language. The words themselves are a delight to learn. Learning them makes you feel that you are off to visit a foreign country. There you will find centrioles, centromeres and telomeres; visit CG islands and alpha microsatellites; find introns, exons, promoter elements and CAAT boxes; calculate LOD scores, Punnett squares and Kilobase lengths; you can come across horrible compound objects such as restriction-fragment-length polymorphisms (with which I have been struggling for the past twenty years), single-stranded conformational polymorphisms, or uniparental heterodisomy. These sound like words chosen at random, but are in fact part of a new language. Having learned it, you can understand how the complexity of biological systems answers to your classification and categorisation. You find things still happening that are outside your vocabulary. They remain intransigent until someone discovers their secret. Then a new set of words suddenly appears on the scene, and their truth can never be fully known until tested in the fire of experimentation and publication.

There is no such thing as a crash course in this subject. The enthusiastic reader will have to read one of the many books referred to at the end of this book in order to understand more of the details. The catalysts involved, the nature of the different chemical bonds, the other grammatical features of the gene such as the syntax for 'start' signals, 'stop' signals and the 'promoter' regions that tell how the gene is regulated can all be found in one of the many excellent books, such as *Emory's Elements of Medical Genetics* (1995).

The code has a great deal of normal variation, some of which accounts for the individual differences between people such as their eye or hair colour, height or body weight, etc. Some of this normal variation can become useful if the environment changes and an example of this is given in Chapter 5 (i), describing the adaptation of the Norwegian rat to a changing diet containing warfarin. Other variations in the sequence of the code may not have any biological effect, but sometimes the coding sequence seems to go drastically wrong, with alterations in the DNA sequence or miscopying of a gene into the message. Then the cell may make a defective protein, or even no protein at all. This can severely damage the function of the cell and lead to disease.

Why does the code go wrong?

If you are using a set of accurate plans over and over again it is very much in your interest to make sure that such plans are written and stored in a secure and stable form, not necessarily carved in stone or etched on metal, but in something nearly as durable. The plans for life are 'written' in the molecule of DNA. Compared to other substances in the cell, DNA is amazingly stable. After all, our continuation as a species depends in large part on the stability of our

DNA code. You can heat DNA in water up to 65°C, precipitate it with pure alcohol or treat it with weak alkalis and it will not degrade the backbone of the molecule. However its stability is not perfect and errors do creep in. A particularly vulnerable time is when the DNA is being copied for the next generation. It is as though a photocopier makes errors in some of the letters of words that it is copying and these are transmitted onwards to the off-spring. The copying process can sometimes misalign the old and newly forming strand of DNA so that chunks of the new gene are left out of the copied sequence. Chemicals called mutagens can destroy or damage individual bases of the DNA to cause errors in the coding sequence. Substances such as dyes (acridines), ben-zene and mustard gas can all do this. DNA is also sensitive to ultraviolet radiation and X-rays that destroy some types of bond-ing within the molecule. Some of these mutations may have no effect on the function of the gene, whilst others may occasionally even confer an advantage on the activity of the gene. This spontan-eous variability has been the basis of evolutionary change since life began on our planet about four billion years ago. The distribution of different genes in animal populations, whether deleterious or beneficial, is largely regulated by the evolutionary forces of natu-ral selection.

Coding errors (mutations)

How does the code go wrong to produce a mutation and so lead to disease? Consider a set of three-letter words making sense of the sentence 'the cat sat off the mat' where each letter stands for one of the building blocks of DNA (I have included more letters than in DNA just to make sense of the sentence):

THE CAT SAT OFF THE MAT

Suppose that the third letter E of the first word THE is destroyed by some means. The change is called a mutation, and the sen-tence taken again in triplets now reads:

THC ATS ATO FFT HEM AT

This is because there is very little punctuation in the gene sequence, so that if one letter is dropped from a triplet the next letter downstream is used instead. This pulls all the other triplets out of order and the sentence now becomes a meaningless jumble of letters. The protein made from this particular coding sequence would not work properly, or perhaps not even be made at all, and this might severely disrupt the function of the cell.

Now suppose the fifteenth letter E of the original sentence is replaced by a C; the sentence will read:

THE CAT SAT OFF THC MAT

The last triplet MAT is not affected because a new letter C has replaced the old one E. Only one triplet out of the whole sequence is altered by this change and all the others retain their correct order. The sentence makes more sense and may well be read off by the machinery of the cell to make a protein with perhaps just a slight defect. So if this gene was helping to control, say, the level of blood cholesterol, the change in the sequence of letters could act as a marker for the defective gene. This in principle is how markers can be used to predict the future occurrence of disease such as raised blood cholesterol, diabetes or early heart attacks, depending on the site of the mutation and the function of the gene. The change in the letter sequence may not cause the disease *per se* but only make it more likely to occur if unfavourable inter-actions take place with environmental factors or with other genes.

So a mutation can be seen as a change in the coding sequence of a gene that either impairs its function or leads to the manufacture of an abnormal gene product (that is a defective protein). The variation in the sequence of letters does not have to lie within the gene. It can be located on the thread of DNA that intervenes between the genes and can affect the regulation of the activity of that gene. Such mutations are usually close and they can be used as a marker for that particular gene. Other sequence variations can be found in the intervening DNA but probably have no biological effects. They can still be used as markers for nearby genes. These facts about genes and how they code for proteins contain no particular morals or sermons, but the combination of these facts with the newly developed reproductive technologies has opened the way to creating the 'designer' baby, and here ethics come very much to the fore.

3

Designing Babies

Dear God, no! That old-fashioned way's the least
Efficient and a great deal too much trouble.
That sort of thing's all very well for animals.
Mankind needs something nobler, that suits
His nobler nature. What you see is Science's solution
Of Nature's greatest mystery – Evolution.

—Goethe, *Faust*, Part II
(Wagner's response to the creation
of a baby in a glass flask)

Chance is a regrettable impurity in the unfolding of events in life. Why leave it to chance as to whom one should marry or live with? Yet until now parents have left entirely to chance which of their genes they pass on to their children. They will if they wish soon be able to handpick which genes are to be transferred – leading to the concept of 'designing a baby'. Other people may prefer to think of it as producing a 'Frankenstein baby'. However, it is already being done to a limited extent in some fertility clinics in the USA where parents can now choose the gender of the child they are expecting. From sex selection to choice of other characteristics such as reduced liability to chronic disease, to stature, or even to hair colour is probably only a small step. This might cost the parents about £10,000 ($15,000) compared to the cost of about £100,000 if the parents were to educate a child at a private school. Newborn babies often begin their life as a loud noise at one end and no sense of responsibility at the other. But it does make sense to give them the best possible start in life by not handing on any of the worst parental genes. A single parent could even be able to provide almost all the genes of the child from him- or herself by the new technique of reproductive cloning.

Our reproduction is considered a great marvel and mystery. Martin Luther (1483–1546), the great German religious reformer, wished that God had continued the generation of our species by fashioning us out of clay. Indeed, the pleasures of the sexual act are momentary, the position ridiculous and the expense often excessive (to paraphrase the Earl of Chesterfield). Whether Luther would have approved of our current ability to create babies, not out of clay, but in test tubes and avoiding much of the preliminary activity is a moot point.

How to 'design' a baby: two crucial steps

1. Test-tube babies

The world's first test-tube baby was Louise Brown, born in 1978. Her birth demonstrated that it was possible to create a human embryo outside the body in a test tube, then replace it in the mother's uterus and see it successfully implant and develop into a child. Two British doctors, Dr Edwards and Dr Steptoe, removed one of several eggs from Mrs Brown's ovary and added Mr Brown's sperm to it in a dish, where fertilisation occurred. After the egg had divided several times it was placed back into Mrs Brown's womb, where it developed successfully into a baby girl.

For readers who want more information about how this was done a flow sheet is shown in the Appendix, Figure 1. To elaborate briefly here, the female eggs are removed by a needle and syringe from the ovaries. They are then mixed in a glass dish with the husband's sperm (or that of the donor if the husband is infertile), aiming to obtain about 100,000 mobile sperm per millilitre of test-tube fluid. One plucky sperm flashes its tail furiously and harpoons itself into a female egg. The embryo that develops from each fertilised egg is incubated in the dish for one to two days at body temperature. Those embryos whose cells appear to be dividing normally are placed into the mother's uterus or into that of a surrogate mother. When possible at least three embryos are

replaced simultaneously since this increases the chance for a successful pregnancy. It is unusual for more than one baby to develop
at a time. The procedure is most commonly done for women
whose ovarian tubes are absent or have been irreparably damaged
by previous disease or surgery.

Initially, the media response was one of horror and disgust.
The process was unnatural and repellent. It was 'playing at God'.
Before the birth dire warnings were issued by top scientists
(including the Nobel prize winners James Watson and Max
Perutz) that such an experimental pregnancy could result in a
very deformed baby because of the uncertain conditions in the test
tube when fertilisation of the egg occurred. It was thought that
embryonic deformities might be produced similar to those resulting from the use of the drug thalidomide by pregnant mothers,
where the limbs of the foetus fail to develop properly. In fact, the
process of pregnancy tends to filter out the defective embryos and
they do not develop to full term. Thalidomide attacks the baby's
development at a much later stage in pregnancy when the embryo
is well established in the mother and so development goes on to
full term even though the limbs may not have developed properly.

After the birth of the next few test-tube babies public reaction
became more hostile, particularly from the pro-life and anti-
abortion groups. Politicians rambled on until their words all but
lost their meaning. Enoch Powell, a Conservative Member of
Parliament at the time, almost managed to get a bill passed making
it unlawful to do any work on human embryos at all other than
implanting them into a specifically nominated mother. This would
have blocked all attempts to improve the methods such as the
freezing and storing of fertilised eggs or early embryos which
might make the procedure more efficient. The response was very
similar to contemporary comments on human cloning.

Now, more than twenty years after the birth of Louise Brown,
many thousands of babies have been conceived by *in vitro* fertilisation (IVF). Fertility clinics using IVF are not confined to the
wealthy countries of the West. By 1994 more than thirty-eight
countries had established IVF clinics including Egypt, Pakistan,

Thailand, Turkey and Venezuela, and more than 150,000 babies have been born around the world using the procedure. This shows how quickly a new technology can develop and expand when society feels a need for it, despite the early strictures and adverse criticisms.

The overwhelming desire of couples to have their own children, couples who will 'try everything' regardless of cost, is made even stronger when some biological quirk prevents them going down the normal channels. This has been the main driving force behind the extraordinary expansion of assisted reproductive techniques. Infertility treatment has become big business in the USA. More than $1 billion is spent annually on these treatments with more than one million people seeking such help annually. Although this increase may partly reflect the decreasing numbers of suitable children available for adoption, it reflects the very real desire for couples to have a genetically related child. In 1987 there were fewer than fifty IVF clinics in the USA; in 1992 there were more than 230. The procedure creates no media reaction nowadays and one wonders if the new reproductive technologies such as designer babies or cloning may not go the same way.

The results of *in vitro* fertilisation are good and the incidence of abnormalities in these test-tube babies appears to be no higher than in the general population of newborn babies. However at present only about one in five embryos that are placed in the uterus actually implants successfully and leads to pregnancy. This failure rate of 80 per cent is still quite high. The best results are consistently obtained in women under thirty-five whose male partners have healthy sperm.

Test-tube fertilisation techniques have triggered another new industry – commercial surrogacy. Women undertake a contract with their clients to 'rent' their wombs out for a period of nine months to grow another couple's baby. Kim Cotton became Britain's first surrogate mother in 1985 and she was paid £6,500 to bear the child. She later sold her story to a tabloid newspaper for £15,000 ($9,375). Subsequently more than 4,000 children have been born in this way, mainly in the USA and the UK.

Test-tube babies are not yet 'designer' babies. One more step has to be undertaken.

2. Selecting the embryo by genetics

Before replacing the test-tube embryo into the mother's uterus it is possible to sample one or two cells when the embryo is composed of about sixteen to thirty-two cells without apparently causing any harm to subsequent development. The DNA from these cells can be extracted and studied for the presence of harmful genetic mutations. If any deleterious genetic variants are found the embryo need not be implanted. Other test-tube embryos can be screened until a 'good' one is found and this is then used to start a pregnancy. This is a bit like the common practice of quarantining sperm from male donors before use for six weeks by freezing them to check the father for such transmissible diseases as hepatitis or AIDS.

The technique of screening the embryo before replacing it into the mother was first established in the 1980s following pressure from married couples who were at risk of transmitting a serious genetic disease carried on the X-chromosome and who had moral objections to abortion if the diagnosis were to be made later in the pregnancy. Such diseases include a premature form of blindness (retinitis pigmentosa), a type of mental retardation and a muscle wasting disease (called Duchenne muscular dystrophy). The method has now been extended to other serious genetic disorders including cystic fibrosis, a rare cancer of the intestine and a fat-storage disease of the brain (Tay-Sachs). Throughout the world more than one hundred babies have been born after sampling cells from the pre-implantation embryo and no increase in abnormalities in the resulting child has been reported. All the variant genes described in Chapters 12 and 13 can be detected by similar methods. The main advantage of this technique is that it avoids the need for abortion of a defective embryo if found later during the pregnancy. The main disadvantage is that it is a labour-intensive procedure and the failure rate is high. It also 'medicalises' the process of conception, requiring multiple hospital visits by the woman for tests and treatments.

A further development that is just over the horizon is the technique of replacing or adding genes to the initial cells that start the embryo on its course of development. This technique is already used extensively in animals, usually mice, by either adding new genes or deleting genes that are already present. This is done to work out the function of the genes that are either inserted into or 'knocked out' of the mouse's genome. The development of the embryo can then be studied to see if the genetic alterations make the mouse more or less likely to develop certain diseases or other features. It has recently been used in non-human primates. The egg cells of rhesus monkeys were injected with jellyfish genes. After test-tube fertilisation the healthiest of the embryos were implanted into surrogate mothers. One of the subsequent baby monkeys was found to carry the jellyfish gene in all its body's cells. Presumably the gene will be found in the monkey's germ-line cells as well and be transmitted to the next generations ever after.

The extension of this technique to human embryonic cells for genetic enhancement is forbidden in most countries but has been used in the USA to transfer mitochondrial genes into the fertilised human egg, correcting possible defects to allow for normal embryonic development. It has also been used successfully to introduce new genes into human blood cells for the treatment of a very rare disease. Children with this disease cannot mount an immune reaction to destroy invading bacteria properly so they have to live in a sheltered and protected environment. The disease (severe combined immune deficiency syndrome) can be caused by many things but one is the deficiency of a particular enzyme called adenosine deaminase. Two children were treated by taking out their white blood cells, inserting the missing gene for the enzyme, and then returning the corrected blood cells back into the children. The first child to be treated made a remarkable and sustained recovery of a whole range of immune functions and is now able to lead a normal life instead of being imprisoned in an artificial germ-free environment. Other children have been successfully treated by gene therapy for variants of this disease.

If this gene can be inserted correctly into human cells why not use many other genes? This would not be such a novelty as it sounds. Contrary to popular belief, genetic modification of people is not particularly modern. It was done in the Dark Ages. Inoculation against smallpox was invented centuries ago in the East. It was the only means of combating this terrible and often fatal scourge. The idea arose from the observation that one attack of smallpox effectively protected the survivor from subsequent, fatal attacks. The procedure was to take fluid from the pox of a mildly infected person and inoculate it into a healthy person in the hope of causing a low-grade infection. The protection occurs because the viral genes provoke the manufacture of proteins that neutralise any further encounters with the smallpox virus.

Why the protection persists for so long is still a mystery. It is possible, but not yet definitely proven, that some of the viral genes become incorporated into the person's nucleus somewhere to maintain the protective response over the years. The person may therefore be genetically modified. Unfortunately the inoculation of the virus did not always lead to a mild form of the disease and sometimes even proved fatal.

Edward Jenner (1749–1823), a country doctor practising in Gloucestershire, refined the technique by using a closely related pox virus transmitted by cows, namely the cowpox virus. He had noticed that milkmaids tended not to contract smallpox. Through his research, he showed that infection with cowpox did indeed protect people from developing smallpox. This was the birth of the modern era of vaccination and similar vaccination has been extended successfully to a variety of other viral diseases such as measles, chicken pox and poliomyelitis. The techniques employed have been so effective as to all but completely eradicate smallpox infections throughout the world.

Genetic modification of a person can result from other naturally occurring infections, as with the herpes virus, which can cause sores on the lips and genitalia. The infection involves incorporation and persistence of the viral genes in the person's own nuclei of the skin around the lips. The virus can remain dormant or be activated by exposure to sunlight or infections to produce the characteristic blisters around the mouth.

So why not insert other genes of our own choice? The problem is complex. First the gene has to be isolated. This is not too difficult, but delivering the gene into the nucleus of the target cell and making sure it is inserted into an appropriate position in the chromosome under proper regulation is still both difficult and inefficient. It would be extremely hazardous to use the technique in human cells during early embryonic development. The hope for the future is to use specially engineered viruses that incorporate the gene of interest to harmlessly 'infect' the nucleus of the target cell and somehow insert the transported gene into its correct position for regulated function.

Poor performance of the infecting virus continues to be the major problem. The transporter virus can be unstable, it can provoke an immune reaction in its own right and be destroyed, it can introduce too many copies of the gene in the wrong places on the chromosomes, and it can interfere with the proper regulation of the gene once it is inserted. The techniques will surely improve, and in time it may become possible to repair or replace defective genes as well as to add genes where those present are not functioning at all.

The improved techniques could eventually be used to modify genetically defective embryonic cells, and also perhaps to add genes to an already normal embryo to enhance its future development. In the short term the method might provide small corrections of naturally occurring genetic defects that impair the lives of so many people. If the technique gains wide public acceptance the number and variety of possible genetic enhancements could rise exponentially. Gene modifications that were once thought impossible will become indispensable, provided there is someone to pay for them.

The timescales of such future developments are difficult to predict and will largely depend on the financial resources available for such work. Although many people may have hoped that no money would be available, by the beginning of 1997 more than 200 human gene delivery experiments had been performed and more than 2,100 patients had received some form of gene transfer. More than 1,700 of these were in the USA. One very promising line of development is to inject haemophiliacs with the gene to replace the missing blood component. One of these is a protein called Factor IX and the gene (attached to the appropriate carrier) can be injected directly into a suitable thigh muscle. The DNA of the gene incorporates into cells and starts manufacturing enough of the clotting Factor IX to improve blood coagulation and may reduce the number of painful episodes of bleeding. By the year 2010 it is likely that 'cutting and pasting' genes into patients will be as widely accepted as organ transplantation is now, and will be as simple to perform as some types of vaccination. The gene to be transferred may either be injected in its viral form under the skin, or even added to a sugar lump to be swallowed, in the same way as the poliomyelitis vaccination.

There have already been some fatal side-effects from such procedures. Jesse Gelsinger, an eighteen-year-old boy, was undergoing treatment for a rare but not fatal disorder that prevents the liver from eliminating ammonia, a toxin produced during the breakdown of proteins (ornithine transcarbamylase deficiency). He had kept the problem under control hitherto by dietary modification. However, staff at the Institute for Gene Therapy at the University of Pennsylvania decided to treat him by gene replacement therapy. As a result of this he died. The treatment triggered a severe immune reaction after the infusion of the replacement genes, resulting in multiple organ failure. The US Food and Drug Administration (FDA) investigated the case and ordered closure of the Institute. They found several irregularities. Paul Gelsinger, the boy's father, said: 'It looked safe, it was presented as safe. And it was going for the benefit of everyone . . . We gave our consent but in no way was it informed.' The FDA criticised the informed consent procedure. It was not properly documented and it was difficult to find out who had conducted or attended the discussions. The Institute was accused of giving Jesse too high a dose of the modified gene preparation, ignoring clinical signs that he was not fit to take part in the test and concealing the fact that previous volunteers had shown toxic reactions to the injection. They also omitted to state that monkeys involved in a similar trial had died. The Institute, which has an annual budget of $25 million, was headed by the distinguished scientist James Wilson, who, as founder of the biotechnology firm Genovo, stood to benefit financially from a successful outcome of the research study that led to Gelsinger's death. Was this a case of conflict of interest? Jesse Gelsinger's death is believed to be the first directly resulting from gene therapy. The episode has sent shock waves through the research institutes around the world dealing with this type of work.

Genetic enhancement, or the addition of genes to human embryonic cells, has only been attempted once. The reasons for this are rather similar to the objections raised against embryo selection. To many people the idea of embryo enhancement or selection is anathema. It goes directly against a proper respect and reverence for the natural processes of conception: keep to the traditional ways of doing things, conserve the old forms. But if traditional sexual reproduction is such a natural phenomenon, why are there so many books written on how to do it?

More serious objections, with their possible counter-arguments, are:

(a) Embryo selection or enhancement is like 'playing God'. It will bring down certain retribution on arrogant mankind. The Rebel Angels fell headlong out of heaven (or were they pushed?) for trying to assume the mantle and power of God. We, in our breathtaking arrogance, having just learned to read the genetic code developed during four billion years of evolution, are already saying: 'Yes, but wait a minute; I think I can improve on this.' One could, however, argue the opposite case that all knowledge comes from God. He gives it to us little by little when we are ready to use it properly. We may make some mistakes in its use, but this will only sharpen our ethical sense and make us revere God all the more.

(b) It is quite immoral to select one embryo instead of another.

The selection of embryos before implantation into the womb raises the same ethical issues as the choice of aborting a foetus once pregnancy has started. Either way, some of the embryos are going to be destroyed. It does not matter what stage of development the embryo has reached at the time of the choice, the ethical issues are the same. It is murder of a defenceless individual. Others would argue that selection before implantation occurs at such an early stage that it is no more immoral than preventing each of the millions of sperms or hundreds of eggs that are produced from developing into an unwanted child. What the technology does is to allow the one or two children that parents do bring into the world to be free from some of the unpleasant diseases or disorders that are commonly found.

(c) There are objections to interfering with the natural processes of conception. A new child should be considered more as a 'gift' than a technological product. The new technology alters our ideas about 'being human' by treating us like a manufactured item, except that if you get the design wrong there is no 'product recall'. But all medical interventions are unnatural. If your heart stops beating most people will accept measures such as a pacemaker to get it restarted, even though it is unnatural. Restoring vision by doing a corneal transplant is unnatural. But the beneficial effects outweigh this and it is now a perfectly respectable procedure.

(d) People with the disorders that are being rejected have protested that they would feel undervalued if medical procedures were developed to prevent the birth of children with their type of disorder. Embryo selection would reinforce the idea that they are undervalued members of society fit only for exclusion or termination. But this is the same as reasoning that people with, for example, paralysed legs due to a previous infection with poliomyelitis should stop children at risk from being vaccinated against polio to prevent paralysis because they as adults would feel undervalued. A disability overcome can become an asset and even turned to good account. Beethoven became deaf; Toulouse-Lautrec was born with a genetic disease of the bone (a rare form of hardening of the bones called osteopetrosis) leading to limb deformities; and Stephen Hawking, the physicist, has a severe muscle wasting disease. Itzhak Perlman, the violinist who is paralysed from the waist down because of poliomyelitis, has said, 'Ask any of us who are disabled what we would like in life and you would be surprised how few would say, "Not to be disabled". We accept our limitations.' But should gene therapy (replacement or enhancement) be viewed any differently from the use of vaccinations to prevent infections?

(e) Embryo enhancement or selection could be hijacked by politicians and used, as in Huxley's novel *Brave New World*, to establish a totalitarian state. Undue pressures can indeed be placed on pregnant women by various institutions. A health insurance company tried to force a woman carrying a child with cystic fibrosis to have a termination, because they were not prepared to pay medical expenses following the birth of an affected child. This is a clear case of influencing choice from a position of financial strength. Life insurance might in future be provided by some companies only for those children who have been genetically screened as embryos so as to exclude serious disease or any unfavourable genes that might incur costs for the company. Regulatory measures will have to be put in place in future to curb this type of coercive discrimination if the public is to accept the technology.

(f) Embryo selection will have long-term consequences for society. At the present time the influence of embryo selection is too minuscule to have any effect on the human gene pool. But if the technique becomes routinely incorporated into our reproductive habits, what consequences might one envisage? This would depend on whether the technique would be freely accessible to everyone irrespective of wealth or whether it would require cash payments from the more affluent members of society. In the latter case, the procedure might cause fundamental splits in society. Each new, wealthy generation would slowly but surely accumulate the benefits and advantages from the previous generation by use of these genetic selection procedures. A poorer sub-class would develop without these benefits and diseases such as diabetes, heart attacks, strokes, cancer and dementias could become more wide-spread among them. Economic class differences could gradually turn into genetic class differences. Any widening of the gap between the 'haves' and 'have-nots' could further destabilise society and lead to greater inequality.

(g) Over the long term could embryo selection reduce the diversity in the human gene pool if we were all to select for roughly the same qualities? Biodiversity is an important factor for any species because it enables it to adapt to a changing environment. For example, the abnormal gene coding for the red pigment of blood (haemoglobin) produces in some people of black African origin the severe condition of sickle cell anaemia when both maternal and paternal genes are defective. But the abnormal gene can provide protection against malaria if only one copy of the bad gene is inherited. This mutation has survival value in malarial countries and it would be no good eradicating the mutation from the human gene pool unless malaria itself had been eradicated. So artificial selection against particular genes supposed to be deleterious may over a long period of time actually eliminate beneficial and protective genes in the process. For African–Americans in the United States there is no benefit to be had from one copy of the gene because malaria is no longer found in North America. The defective gene only has deleterious effects when it increases the chances of a child being born with two copies of the bad gene and so developing sickle cell anaemia.

Everywhere one looks in the biological world there is selection. Evolution is largely driven by natural selection, so why should we not introduce a form of social selection as well? We make a choice of our sexual partners. If there are fertility problems, we can take the option of whether to use egg or sperm donation, or we can choose surrogacy if the female has severe uterine disease. After birth we are constantly using selection procedures for education and choice of teachers, for appointments to jobs or promotions. Embryo selection in its turn is here to stay. It will always be used to some extent. The only question is how far it will go. It may be better to allow complete freedom of reproductive choice as to the methods that are currently available and safe as a defence against too much state interference. Such individual choices are likely to be as diverse as the people making those choices. A tolerant attitude towards this diversity should allow people to follow their own inclinations provided it does no harm to other members of society and, particularly, no harm to the unborn child.

In some cases embryo selection involves life and death decisions where there appears to be no obvious guiding morality. In 1987 Angela Carder was terminally ill with leukaemia in the George Washington University Hospital in Washington, DC. She was also pregnant. Initially she was prepared to sacrifice her own comfort for the wellbeing of her unborn child, on the understanding that the baby might have a chance of survival if the pregnancy continued for another few weeks. In the end the doctors concluded that the child's chances of survival were remote unless Mrs Carder underwent a Caesarean section. She refused to do this. The hospital's response to the dying woman's express refusal was to obtain a court order forcing the Caesarean section upon her. Unfortunately neither she nor her child survived. A subsequent appeal led to the verdict that such an intervention should not have taken place (she had the right to refuse medical treatment), but the willingness of the first judge to authorise surgery shows the extent to which the interests of the unborn child took priority over the rights of the mother.

Whether or not one approves of Angela Carder's refusal to act in the best interests of her unborn child, society's response to the protection of the unborn child appeared both clumsy and unjust. The conflict is between an individual's freedom of choice and the notion that people should be helped or even forced to do what is deemed to be in their best interests. There appears to be no 'right' judgement in this particular case. It just depends on which point of view you take – the unborn child's right to life or the mother's right to make decisions about her own pregnancy. One problem is that the people doing the 'deeming', even if well intentioned, are laying claim to knowledge of outcomes that they do not – and cannot – possibly ever possess. This particular problem is of course not directly related to embryo selection but one can foresee similar issues arising where the rights of parents to start a pregnancy (involving embryo screening, enhancement or cloning) come into direct conflict with society's current attitudes and prejudices that appear unfair or heavy handed. It appears more liberal to allow individuals the freedom to make their own choices provided it is not in direct conflict with the rights of others.

Is embryo selection eugenics?

Ethics is a matter of comprehension. The use of embryo techniques to select genes that affect personality traits such as obesity, alcoholism, aggressive social behaviour, neuroticism or homosexuality, as described in Chapter 13, have to come under careful scrutiny. This is where the 'designer' baby becomes a reality. Parents want, and perhaps have the right, to choose the best possible health and education for their children. If parents have the right to spend large sums of money on their child's education at a private school, why should they not be allowed to spend an equivalent amount on ensuring optimal gene transfer to their children at the start of life? Genes and environment are the critical issues affecting a child's future achievements and success in life. Why leave the transfer of genes to chance, and yet take infinite pains to control the education and environmental health of the child? If parents want and are prepared to pay for gene transfer and enhancement techniques it is difficult to put forward good reasons to stop them provided the methods are safe. A prevailing social view has been that an individual's freedom is paramount provided his or her behaviour harms no one else. They should have the freedom to choose whatever technology is available provided it is safe. This of course does not necessarily mean that they are making 'good' choices; it all depends on what the values, morals and beliefs are that govern their choice.

It is easy to imagine that the value system of the privileged middle classes concerning their children constitutes some sort of divine wisdom, but this is not necessarily so. Are their choices made from the specious vanity of having a 'perfect' child? Or from a real belief that their value systems should be transmitted to the future child whether it be for personality attributes, stature, eye colour or whatever? After the simple choice of the sex of their child, parents could go on to select an embryo with none of the known genetic variants that predispose to personality traits that they may consider undesirable. The parents may not want to pass on their 'best' genes, however these may be defined. But they may at least want to avoid passing on their deleterious genes that are much easier to identify.

A widely publicised case in 2000 of 'designing' a baby was that of the Nash family from St Paul, Minnesota, USA. They already had a daughter, Molly, who suffered from a rare genetic type of anaemia (Fanconi's anaemia). They decided to have a second child, one who they hoped would not carry the anaemia gene but would be as compatible as possible with Molly. Cells from the new designer baby would, it was hoped, be transplanted into Molly and so provide a cure for her anaemia. This was done by creating twelve embryos by standard test-tube techniques and then examining single cells from each of them to see which would be the best fit for Molly. The most suitable embryo was then selected for implantation into Mrs Nash's uterus; Mrs Nash gave birth to a boy, called Adam. Stem cells from Adam were then transplanted into his sister Molly in the hope that they would develop into marrow cells to make the blood cells required to cure Molly's anaemia. There would be a 90 per cent chance of curing Molly using this method. In fact, Molly's blood count has steadily increased since the transplant, indicating recovery of the bone marrow. There is greatly increased cellularity of her bone marrow that is making, for the first time in years, a plentiful supply of blood cells. This would count as a cure for Molly's condition.

By designing a new life to save an existing life, Mrs Nash hoped to end up with two healthy children. This she did. It is difficult to see it as an ethical problem. However, committees of the great and the good are debating the ethical issues of this sort of procedure. Rightly they warn of the short steps needed for a more widespread use of pre-implantation genetic screening. Where will it end? The next steps along the line are easy to predict. Designing a baby to provide a kidney for a brother or sister; then creating a baby to provide a liver transplant for a sibling; and then perhaps because there are too many children in the family the new child may be selected to carry a 'suicide' gene that will cause the child to die young after donating the liver to his sibling.

In future doctors might say to a prospective mother: 'I recommend using embryo 15 for your next baby which is likely to produce a child of above average intelligence, an excellent immune system making the child resistant to all sorts of infections, and also having a low risk of developing early onset of cancer, heart disease, or Alzheimer's dementia as an adult.' Which parents would refuse such an offer? How much notice would they pay to ethicists, philosophers or governmental agencies who pontificate on the dangers of choosing 'designer' babies from a range of desirable genetic traits? It appears likely that it will be the parents who will be the driving force behind adoption of these techniques when they encounter problems with their own families. They may find the rules and regulations imposed on them by outside authorities to be largely irrelevant. If the procedure is safe but nonetheless outlawed, it could lead to a clandestine and unsatisfactory service being set up to provide for the needs of the afflicted families, as happened with illegal abortions in the last century.

The procedure of 'designing' babies by enhancement and selection techniques would fall within the original definition of eugenics, which in the early part of the twentieth century used selective breeding and sterilisation as the only methods at its disposal. The subject is potentially open to all the abuses of discrimination and suppression by authoritarian governments that followed in the train of the eugenics field in the first half of the twentieth century.

Expected take-up of eugenic technology

What might individual parents, rather than governments, actually want from the methods employed in embryo selection and enhancement to help them get a 'better' baby? One can form some idea of this from the case of genetically engineered human growth hormone.

In the 1980s, two major American companies, Genentech and Eli Lilly, were awarded patents to market a new genetically engineered growth hormone for the treatment of a few thousand children suffering from dwarfism in the UK and USA. Since the market share for the hormone was expected to be so small, both companies were awarded 'orphan' status for their drug. With this went several privileges to compensate both companies for making an investment in a drug with such a small sales potential.

Surprisingly, by the early 1990s the genetically engineered growth hormone was doing extremely well and the two companies shared a market of nearly $500 million in sales in the USA. It turned out that the growth hormone was being prescribed by doctors to treat children of short stature but who were in no way medically deficient in growth hormone. Some parents believed that taller people had potentially greater advantages in life than those of shorter stature who were less advantaged, a discrepancy that would lead to behavioural and emotional problems in later life. Such parents were prepared to pay for the daily injections that would add a few extra inches of height to their children. Doctors who believed that one of the goals of medicine is to optimise the sense of physical and mental wellbeing of people (rather than the prevention or cure of identifiable disease) went along with this. They were aided and abetted by the aggressive promotional policies of both Genentech and Eli Lilly who tried to label children of short stature as being 'diseased' and in need of hormone therapy. Even the National Institutes of Health in Bethesda, Maryland, went some way towards encouraging this by collaborating in a seven-year research project with Eli Lilly to investigate the effects of growth hormone therapy on adults who had been of short stature as children. It turned out that the 'treatment' only added a few extra inches to the final height of the child; today the treatment is less strongly advocated. But the surprising sales of growth hormone give some idea of what to expect for the future of genetic enhancement techniques other than for medical uses.

Another group of people who inject themselves with growth hormone for enhancement are world-class athletes: in particular cyclists, swimmers and body-builders. Winning an Olympic gold medal or winning the Tour de France means financial security for life. Coming second gives very little return. Any slight advantage they can get over their rivals tempts the athletes, especially when the difference between winning or coming second is measured in fractions of a second. Growth hormone injections certainly make the muscles look bigger and stronger, but two scientific studies have shown no actual improvement in athletic performance. It is more the dream of success that leads them on. The extent of the use of growth hormone is uncertain. Large hauls of the hormone have been found with athletes and their accomplices and there are more than 1,000 Internet sites which explain the use of growth hormone for athletes.

The problem this gives rise to is the 'medicalisation' of conditions (stature, muscle strength, etc.) not normally treated as diseases by doctors. It is difficult not to forecast an equally dramatic boom in the techniques used for genetic enhancement when they eventually appear on the market; they may even generate the same enormous profits as cosmetic surgery. If parents are allowed to teach their children their own values, guide them in their religious beliefs or expect them to learn to play a musical instrument, why should they not also choose which of their genetic traits to pass on to their children to enhance their performance or well-being?

Many people consider that the techniques for producing test-tube babies and screening embryos before implantation into women should be regulated by non-commercial agencies in order to combat the push of the marketplace and the pull of unreasonable consumer desires. After the birth of Louise Brown, Britain set up a voluntary system of regulatory clinics undertaking these procedures. In 1990 Parliament set up a special body, the Human Fertilisation and Embryology Authority, to regulate the field by law. Without the acquisition of a licence from the Human Fertilisation and Embryology Authority, no human embryo can be created outside the human body, nor can eggs or sperm be stored for future use. Human embryos up to the age of fourteen days can be created for screening or research purposes, but an additional licence is required from the same authority if a screened embryo is to be replaced into a woman's uterus.

The next logical stage after gene replacement or enhancement therapy is to give the child a defined set of genes that you know works – namely your own. To create a child in your own image is to embark on the process of cloning.

4

Cloning Babies

A joyful mother of two goodly sons;
And, which was most strange, the one so like the other
As could not be distinguished but by names.

—William Shakespeare, *The Comedy of Errors*

A unique event in the history of the reproduction of mammals occurred in 1997 when a nucleus of a cell from the udder of a sheep was extracted and transferred into the egg cell of a sheep from which the nucleus had been previously removed. The new egg cell was then implanted into the uterus of another sheep which led to the birth of perhaps the most famous lamb of the century, Dolly. A flow sheet of the experiment is shown in the Appendix, Figure 2. The technique of 'cloning' has been used before in other vertebrates such as frogs in 1952 as well as in sheep in 1996, but always using nuclei derived from embryonic cells. This was a sort of artificial twinning process which enabled scientists to produce multiple copies of the same individual if the original nuclei were all taken from the same embryo. What was different now was the demonstration that an adult nucleus contains all the genes in working order to programme an egg cell which can develop into an adult. It was previously thought that the nucleus of adult mammals somehow had most of its genes switched off during development and could not revert back to instructing an egg cell how to develop into an adult from scratch. In the 'Dolly' experiment the adult cell had to be cultured under special conditions before extracting the nucleus for transfer, otherwise the embryo would not develop. It should also be noted that Dolly was the result of 277 attempts to fuse an adult nucleus with an enucleated egg, and of these only twenty-seven embryos developed normally for the first week and only one developed into a fully grown sheep.

The technique of cloning is, therefore, not very efficient at present but would be expected to improve with more research. The experiment showed for the first time that asexual reproduction in mammals was feasible, and that it is only a matter of time before the method is used to give birth to a human clone, or 'twin', separated by one generation. The twin would not be absolutely identical to the adult donor. Although the nuclear genes would be the same as those of the donor, there are also some genes in other cellular structures called mitochondria that would be supplied by the egg cell coming from the female.

The report of reproductive cloning of Dolly the sheep in the journal *Nature* in 1997 created a furore – mainly the gut reactions of those opposed to the technique ever being used in humans. Some people felt that there are limits beyond which we should not go. It appeared to raise fundamental questions about the nature of human individuality and how a human life should begin. Although the technique used is no different in principle from test-tube conception, whereby the nucleus of a sperm is now substituted for the nucleus of an adult cell, the public outcry from those in all walks of life was immediate and almost instinctive: the use of adult nuclei for cloning humans should be made a criminal offence. The Director of the World Health Organisation stated: 'The WHO considers the use of cloning for the replication of human individuals to be ethically unacceptable.' The Director of UNESCO said, 'Human beings must not be cloned under any circumstances.' The European Parliament passed a resolution stating that: 'the cloning of human beings . . . cannot under any circumstances be justified or tolerated by any society because it is a serious violation of fundamental human rights'. These are grandiose expressions of course, but unfortunately they do not seem to mean much. The US Federal Funding Committee banned financial support for research into cloning of human embryos from 1998 for the following five years. The US Senate narrowly failed (fifty-two to forty-six votes) to pass a bill whereby the cloning of human embryos would constitute a criminal offence.

In 2000 the Japanese government introduced legislation to outlaw the transplantation of cloned human embryos into a uterus, with severe penalties imposed on anyone breaking the law. The spectre of political megalomaniacs – the Hitlers and Saddam Husseins of this world – reproducing themselves almost exactly without an ameliorating mixture of genes from a woman was almost too frightful to consider. The film *The Boys from Brazil* no doubt added fuel to these fears. Likewise the Dutch government has prohibited all work on human cloning, therapeutic or otherwise, and for good measure has banned gender selection and scientific research on human embryos for the next five years. This ban goes against their own Netherlands Health Council, who did recommend limited exploration of scientific research on human embryos, but it is said that public opinion in the Netherlands considers this to be unacceptable.

To try to enforce measures without adequate public debate naturally leads to disputes. To many in the scientific community it was almost equally repugnant to ban by law a scientific enquiry into any fascinating area of research, especially when situations could be envisaged where cloning could have beneficial consequences. Couples with intractable infertility, or partners who both carry lethal genes that lead to repetitive miscarriages might prefer to have a genetically related child by cloning rather than adopting one. The unease over the ban in various countries was heightened by the arguments put forward by the various political bodies. Scientists were to be imprisoned by a set of specious and high-sounding phrases: cloning 'violated respect for the dignity of human beings'; it is 'contrary to human integrity and morality'; it violates 'the security of human genetic material' and fails 'to preserve the human genome as a common heritage of humanity'. The European Parliament, apart from stating that cloning humans is a serious violation of fundamental human rights, also considers that cloning is 'contrary to the principle of equality of human beings . . . it offends against human dignity and it requires experimentation on humans'. It certainly does raise the problem about the nature of self-identity. But this has always been a metaphysical quagmire.

All the above statements would appear to apply equally well to other forms of multiple pregnancies induced by artificial means such as *in vitro* fertilisation and twins arising by deliberate embryo splitting. The major difference now is that the 'twinning' is between generations. It is also difficult to understand how the factor of 'between generations' could invoke such unanimous expressions of fear and disgust from political and other institutions. Of course, there are serious implications with regard to the change in family relationships that cloning would produce. If the child were to be cloned from a man, the child's social grandmother could also claim to be the child's mother since she has contributed half the genes.

Before being carried away by such social arguments let us consider two separate issues: cloning for therapeutic purposes and cloning for reproducing a human being.

Therapeutic cloning

The aim here is to prevent the cloned embryo from developing into a child. Instead development would be arrested at an early stage, preferably within fourteen days. Then the cells derived from the embryo would be taken and grown in artificial cultures. The most useful cells to harvest from the embryo would be the stem cells or 'master' cells that have the potential to transform into many different organs. These cells could be of immense use to the adult who donated the nucleus.

The problem with transplantation of organs from one individual to another is that of rejection, whereby the host treats the transplanted tissue as foreign and reacts destructively against it. However, cells from a cloned embryo derived from an adult nucleus used for the transfer would be so similar to the adult that they would not be rejected and would allow successful transplantation of many tissues. Replacement of skin after severe burns; replacement of nerve cells in the treatment of Parkinson's disease; the attempts to restore the loss of brain cells in the dementias such as Huntington's disease; the curing of some cancers by replacing healthy bone marrow in the leukaemias – these are all real possibilities.

The initial objections to human cloning have softened in view of these possible medical benefits. In the UK, the Human Fertilisation and Embryology Authority has recommended that cloning techniques of embryos less than fourteen days old could be used to develop cells or organs for transplantation purposes. With American Congress banning Federal-funded research into the human embryo, this has channelled the research into the private sector. Several American companies, such as the Geron Corporation in San Francisco and Advanced Cell Technology in Worcester, Massachusetts, have already set up production lines for the cloning of embryonic stem cells to provide a variety of tissues for therapeutic and transplant purposes.

Social surveys in the UK to gauge public opinion have produced varying results. In one study conducted by the Wellcome Trust in London, seventy-nine subjects were recruited to represent a cross-section of the public with no specialised knowledge of genetics or cloning, and asked for their opinions. After attending workshops explaining the techniques involved most participants were initially sympathetic to the idea of cloning for therapeutic purposes. On further consideration of the implications others raised strong objections. To some people an embryo at or before fourteen days is not just a blob of cells but is already a person or at least has the potential to become a person, so they would not dream of taking parts from it for their own use and then destroying it. One man thought 'it could be psychologically disastrous if you created an embryo to make a part for yourself and then just destroyed it'. This would amount to using a human embryo as a commodity to supply spare parts for adults, which is quite different from the previous research on embryos. It goes against Immanuel Kant's famous moral dictum that we must not treat another person as a means to an end (in this case for supplying body parts), but always as an end in themselves. It could be considered worse than abortion or judicial murder. At least in the latter two cases we have no vested self-interest in the spare parts that may arise from taking a human life.

Other people have considered stem cell research to be ethically no different from removing organs from dead people for the purposes of transplant surgery. In the UK, Mary Warnock and the committee she chaired in 1984 concluded that the potential benefits of embryonic research for medical purposes outweighed the individual rights of an early human embryo to life and limb. To the Warnock Committee an early human embryo is very much like any other animal embryo at the stage when it is to be used for research. It may just perhaps be entitled to greater respect than those of other species.

Because of the Warnock Report, permission to experiment on human embryos up to the age of fourteen days was incorporated into the Human Fertilisation and Embryology Act of 1990 (UK). This was extended in December 2000 by the UK Parliament who voted decisively to allow research to go ahead for the production of embryonic stem cells to develop organs for transplant, such as bone marrow, skin and nervous tissue. Admittedly this will lead to the cloning of the patient's genetic material to obtain tissues that match perfectly.

Opposition was voiced mainly on the grounds that this represents the first steps towards reproductive cloning. The UK legislation will also be difficult to square with the recent European Parliament resolution that voted (on 7 September 2000) to reject the use of human embryonic cloning for any type of research and that criminal penalties should be instituted for any breach of this. For the European parliamentary members, therapeutic cloning 'irreversibly crosses a boundary in research norms and is contrary to public policy' (whatever that means).

With regard to spare-part surgery, Dolly the lamb is not the only animal to have been cloned. Since 1997 the ranks of cloned animals have swollen to include mice, rabbits and cattle. Another very interesting step forward was the recent announcement of the birth of five cloned piglets. The aim here is to develop an ideal animal as an organ donor for patients requiring spare parts such as kidneys or hearts. The pig is a good animal for this. Their organs are about the same size as those of humans, allowing them to be easily inserted into the human organ bed. The big problem is, of course, rejection of the transplanted tissue. If you transplanted a pig heart into a human it would turn a mottled colour and be rejected within hours. The human sees the pig heart as foreign material, and goes on to attack and destroy it. The pig heart 'looks' foreign to human tissues because of a group of special sugar-linked molecules on the surface of the pig cells. If you could knock out the genes making these sugar molecules then perhaps the transplanted heart would not be seen as foreign and would be accepted by the host. An unlimited supply of human organs for transplant could be in the offing. No more interminable waiting for the correct kidney or for a heart donor to die accidentally and supply the organ that matches your own tissues. It sounds far-fetched but could be in operation within five years from now, provided the problem of transferring contaminating pig viruses in the transplanted organ is solved.

Another therapeutic option for cloning would be to introduce a human gene into the animal nucleus that is used for the transfer and fertilisation of an animal egg cell. One would expect that the adult animal would then make the human protein from the inserted human gene and that this protein could be harvested for medical use. This has been done with six cloned calves created from cells cultured *in vitro* for up to three months. During this time it was possible to insert human genes into the cell nucleus used for cloning. An animal has been produced in this way that secretes a human protein (Factor VIII) that is involved in blood clotting. If people, such as haemophiliacs, lack this factor they can be provided with it after extraction from the blood or even from the milk of the cloned animal.

Reproductive cloning

As discussed earlier, people's instinctive reactions to human cloning were ones of horror and disgust. It cut across a fundamental view that a child is the creation of two separate adults, one male, the other female; and that tampering with the process goes against one of the basic laws of Nature. Nature is that lovely lady who has also given us smallpox, leprosy, syphilis, tuberculosis and cancer, diseases for which we accept intervention without a murmur. But the official views on cloning have been reinforced by public surveys. Those who took part in the Wellcome Trust survey discussed earlier (see page 60) considered almost unanimously that having a cloned child was unnatural, unacceptable, demeaning as regards the rights of the unique and personal identity of the cloned child, and deliberately deprived the child of normal parenting.

They further predicted psychological damage of the cloned child, including resentment at having its genetic constitution predetermined by a single parent without its consent; resentment at having been conceived as a means of benefiting one individual rather than being in the child's best interest; and eventual confusion about personal identity and family relationships. There is no way to calculate such potential harm – it is purely guesswork. Much of this critique could also apply to a child born by natural means – just read this paragraph substituting two parents for one and it still makes sense.

It is also important to distinguish genuine moral concerns from strong gut feelings of revulsion for a process considered to be unnatural. One could argue that all human intervention, such as surgery for an inflamed appendix, is unnatural and therefore should not occur. Indeed there were strong 'moral' objections to corneal transplants to restore vision when they were first introduced. It was the start, so it was said, of a slippery slope which would lead to the creation of transplant monsters similar to Frankenstein's monster. Now almost every major hospital in Europe has its transplant surgeons; so far the consequences have not been dire – there are no monsters.

Adopting a more pragmatic approach, one could argue that if the benefits of cloning outweigh the risks it might be worth doing, provided all the right safeguards were put in place. Women who have difficulty in conceiving may well consider cloning as a means of starting a family when all the other measures have failed. In the previously mentioned Wellcome survey one woman had been trying for nineteen years to bear a child. She said, 'if that [cloning] was the only way to have a child – it's selfish – but it would be great'. Another woman rejected the use of cloning for herself but felt that it could be acceptable in cases where a woman was 'desperate' and all other methods of conception had failed. A conflict of rights emerges here. The right to have a family is part of the United Nations Charter of 1948, whereas the creation of a child by nuclear transfer methods has been legally banned in Europe, Japan and the USA. After considering all the arguments for and against cloning, someone will eventually have to decide which policy to adopt. And this is the problem. Who will have the authority to make the decision? Will it be public opinion, government committees, professional groups such as doctors or geneticists, or will it be left to individual choice in a free marketplace driven partly by commercial interest?

There are still some fundamental objections to cloning a child that come before all the moral arguments. There is the major question of safety for the infant. An adult nucleus is not the same as an embryonic nucleus. During adult life the nucleus accumulates a large number of changes in the genetic code called somatic mutations. Every time the cell divides there is a chance that an error will occur during the copying process of the genetic code. Many of these errors, the somatic mutations, will occur in genes that are not being used in the adult and will have no obvious effect. But they may be required during the development of the embryo.

By the time maturity has been reached and then during the ageing process it has been calculated that about a quarter of the 35,000 genes in the nucleus have accumulated errors in their code. This happens to a much lesser extent in the nuclei of cells that go to the formation of sperm or egg, since fewer cell divisions are needed to produce these specialised germ cells. Such error rates can be accelerated in different cell types depending on, for example, exposure to sunlight, to X-rays or to particular chemicals or foodstuffs. In sexual reproduction, as opposed to asexual methods such as cloning, the fertilised egg receives a new random assortment of genes from both parents, thus diluting the errors in the genetic code that each parent contributes. If, unluckily, both parents are closely related, perhaps first cousins, they are more likely to possess the same errors in a particular gene; in such a case the 'dilution' effect no longer occurs and their resulting children will be more prone to developing a genetic disease. Hence the warning against first-cousin marriages and even more so against brother/sister marriages. Reproduction by cloning bypasses this dilution effect. All the errors of the adult nucleus used for cloning will be transmitted to the embryo where the ongoing accumulation of errors will add to those already present. This may lead to an increase in the occurrence of diseases that relate to somatic mutations such as the cancers, arthritis and immune deficiency diseases. It would be outrageous to handicap a child from birth with a multitude of genetic errors and subsequent deformities due to the deteriorated condition of the adult nucleus used for transfer.

This Catch-22 situation may very well block the use of cloning children for ever. The extent of expression of these genetic errors after introduction into the egg cell will not be known until it is tried – and it is clearly unethical to try the experiment if there is a risk of seriously harming the newborn child. (It should be noted that Dolly is already showing signs of premature ageing in her chromosomes. Erosions are starting to appear at the tips of her chromosomes that do not normally appear until an animal is much older.)

French agricultural scientists have reported other health hazards arising from cloning. A calf was cloned from the nuclei taken from the skin of the ear of an adult cow. The calf developed normally after birth for the first month but then suddenly stopped making red or white blood cells and died from a severe anaemia at seven weeks. The cloning process seemed to have interfered with the normal development of the spleen, thymus and lymphoid tissue, which led to the failure of the blood components. Doubts are clearly cast over the practical safety of cloning that should take precedence over all other issues.

Cloning a child for the first time would be an experiment and should be subject to the standard principles of research ethics as laid down in the Helsinki Declaration of 1964 (see page 140 for details). The unborn child obviously cannot consent to the experiment, yet the calculated risks for the first cloned child (accumulation of somatic mutations) are large enough to be inconsistent with the best interests of the child's health. Of course, no one has even given consent for their own birth, but their creation was at least not the result of a calculated scientific experiment where man-made hazards can be anticipated.

In spite of the many arguments against reproductive cloning, from the dehumanising aspects, to the child's personal identity problems, to problems of injury to the infant, it must be admitted that reproductive cloning will probably occur somewhere in the world within the next few decades. Our somewhat elastic philosophy of today means that if we know how to do something, someone somewhere will do it. It will not be because the majority approve but because a childless minority acts on it. That is what happened with test-tube babies. Society did not decide to allow it to happen; it was pushed through by childless couples who were desperate to have children of their own. Cloning could even be used in the technique of designer babies by inserting additional genes into the adult nucleus before it is used for transfer to the recipient egg cell.

Within a few months of the announcement of the birth of Dolly, a Chicago scientist, Dr Richard Seed, planned to open a cloning clinic in the USA. Babies could be made using women only; the man now became unnecessary and superfluous. (The flow sheet for this is shown in the Appendix, Figure 3.)

Although Congress had declared a moratorium on research into human cloning for five years, this did not affect the type of private work that Dr Seed proposed. Dr Seed believed that the opposition to cloning would pass in time: 'Gradually the ethical positions will change when there are half a dozen bouncing baby clones. Any new technology creates fear and horror . . . but eventually receives enthusiastic endorsement . . . and that's what I think will happen with human cloning,' he stated in a television interview. He added that if he were banned from operating his clinic in the USA he would go to another country. He had already approached officials in Mexico. There were already four infertile couples on his books for treatment that he hoped to get started within eighteen months. Thus do novelty and dreams get exalted at the expense of common sense.

Dr Seed has views that run counter to the majority of opinions, but few doubt that he has the ability to carry out his plans. Any laboratory that can do *in vitro* fertilisation and pre-implantation genetic analysis can also tool up to perform reproductive cloning. It is a 'low'-technology procedure, quite unlike the know-how needed to make, for example, a hydrogen bomb. A man's reach should exceed his grasp but Dr Seed is perhaps going too far. Would his activity constitute an offence against our sense of natural justice rather than just infringing the arbitrary rules of our present-day society?

David Hume (1711–76), the great Scottish philosopher of the Enlightenment, had a passion for justice but did not recognise such a category as 'natural justice'. He believed that all the rules of justice are basically artificial, being determined by education and the conventions of society. However, the activity – almost an instinct – to care for and protect one's children appears to be a universal feature of all human societies, with almost no exceptions (except for child exposure in Sparta, as described on page 18). Indeed it holds for most mammalian species. It is said that if you spare the rod you may spoil the child, but wanton cruelty or neglect of children provokes great hostility towards the perpetrator. If anything could be called 'natural justice' the customs concerning childcare could be so considered. They are unchangeable by human laws. If they were to be reversed the society doing so would not survive – we have to provide the best we can for the next generation. Loading a child with unnecessary adult mutations, which cloning may well do, could be considered by some people as a violation of natural justice.

5

Three Warnings from History

(I) Darwin and the 'Struggle for Life'

I have called this principle, by which each slight variation,
if useful, is preserved, by the term Natural Selection.

—Charles Darwin, *The Origin of Species*

This chapter, divided into three sections, will trace the history of ideas about eugenics starting from the mid-nineteenth century. They are written with Winston Churchill's adage in mind: 'The further backward you look, the further forward you can see.' One function of the past is to prepare for the future, by learning from the blunders of history. The trouble is that political beliefs like Bolshevism and events such as Anschluss or Guernica, to which the poets of the Thirties so thrilled, are now nothing more than boring topics discussed at school, the subjects of multiple choice exams. It is as if the Berlin Wall was never anything more than a fragment of brick on someone's shelf. We ignore history at our peril. So it is with the history of ideas about eugenics. Consideration of our past abysmal mistakes should teach us at the very least that concern for individual rights over and above the welfare of the state must be at the heart of whatever policies or strategies we adopt for the use of the newly gained knowledge of human genetics. There is no better way than a study of the history of eugenics to provide this perspective.

Everyone eventually has to come round to Darwin. His thoughts have been the measure of the biological universe. He was the first to give a rational and verifiable explanation for the origin and diversity of the vast numbers of animal and plant species that inhabit our planet. He has exerted a greater influence on biology than any other individual either before or after him. This is primarily due to the intrinsic vitality and applicability of his ideas on evolution that have had such a tremendous impact not only on biology but in the fields of philosophy, theology, sociology, ethics and literature, to name just a few.

As Darwin implied, we are not created in God's image. We are descended by evolution from a primitive ape, which was also the ancestor of our nearest relative, the chimpanzee. It is natural selection, whereby each slight variation if useful has been pre-served, and not divine guidance that has determined our course. God has been diminished; he is made to sit astride a pipette or stalk disconsolately around the test tubes of an experiment per-formed by the molecular biologist (which in many cases will come to nothing or at best produce a new piece of unnecessary jargon).

The doctrine of evolution has now been accepted almost every-where. In 1925 the Tennessee Legislature banned the teaching of evolutionary theory in high schools and universities because it contradicted the biblical account of creation – so much do great ideas and small minds go ill together. In that same year a school-teacher named John T. Scopes defied the law and was accused of contravening the Butler Act forbidding evolutionary teaching. At his trial – it was called the Monkey Trial – Scopes was convicted and fined $100. In 1968 the United States Supreme Court finally ruled that anti-evolution laws were unconstitutional. To get around this, legislation in Tennessee now prohibits the teaching of evolution as a 'fact'. It must only be taught as a 'theory'. In the summer of 1999, the school boards of the State of Kansas voted six to four in favour of removing evolutionary theory altogether from the science syllabus in high schools. Religious conservatives making up half of the school board of ten members argued that

Darwin's theory has not been proved. It is based on speculation; it has never been directly observed. It also flatly contradicts the biblical account of creation. This is a real triumph for the creationists, not formally banning evolution (that would be unconstitutional) but simply deleting it from the syllabus.

This move has naturally been condemned by the heads of all six state universities in Kansas who no doubt feel shame and embarrassment that one of the great achievements of human discovery and a central organising principle of all biology is not going to be taught in Kansas. They wrote to the school boards, warning that exclusion of evolution from the school curriculum would 'set Kansas back a century'. However, according to 1999 Gallup polls in the USA only 10 per cent of Americans say that they hold an evolutionary view of the world, whereas 44 per cent believe in strict biblical creationism. Gary Boer, the standard-bearer of Reaganite Republicanism, refuses to teach his children that they are 'descended from apes' and says, 'I just reject the basic tenet of that theory of evolution . . . and so do most Americans'. In the same Gallup poll four million respondents believed that they have been abducted by aliens – so there are some strongly held rival theories as to our origins that are peculiar to the USA.

The Christian Church has also found it difficult to come to terms with Darwin's theory of evolution. The Vatican leaves open the question of the evolution of man's body provided it is believed that evolution takes place under the dispensation of a divine providence and not the chance forces of natural selection. Some people, particularly women, are not so impressed by man's body anyway. They often hold an opposite view, that man could actually be the missing link between the anthropoid apes and civilised human beings. Yes – God did indeed create man, making him arrogant, vulgar and crude, but then realised Her mistake.

Who then does believe in evolution? Anyone who has seen and been convinced by the evidence. There is an almost complete fossil record available for the evolution of some animal species, such as the horse; and there is substantial, but admittedly incomplete, fossil record for transitional forms of *Homo sapiens*. Even more persuasive is the astonishing similarity between the basic chemistry of single-celled organisms such as bacteria right up to the cellular chemistry of humans described in Chapter 2. That means you can insert human genes into the simplest of life forms, such as bacteria, and the human genes will work there using the chemical machinery of the bacterium. The code of life appears to work in much the same way through all living organisms, equally in bacteria and humans. This provides major evidence for a single origin of life on our planet. It would be a very unlikely coincidence if the code had been formed more than once in exactly the same way. It is much more reasonable to believe that all forms of life on earth have been formed from a common ancestral prototype that has passed on its basic biochemistry, including the genetic code, to all subsequent descendants.

Darwin's theory of evolution succeeded in putting the whole of past life into every aspect of every form of present life. From the Darwinian point of view, evolving as a species is more important than actually living as an individual. Darwin demonstrated that the incredible variety of life, with all its complexity and puzzling relations to its environment, could be explained in scientific terms. He also threw down a challenge to theologians and his doctrine of evolution made many theological matters, including the origins of *Homo sapiens*, much more a matter of degree. The evolutionists upset the theologians so much because it was the first time that the Bible could be shown to be conclusively wrong in principle. Other passages in the Bible causing dispute could still be resolved by assuming an all-powerful God working miracles. Thus in Joshua 10:13, when 'the sun stood still and the moon halted . . . And the sun stayed in mid-heaven for almost a whole day', God could have worked miracles to protect the Earth from massive tides and overheating of the Earth's side that faced the sun. But evolutionary theory directly contradicts the account of creation in Genesis. Some creationists tried to win the argument by claiming that the fossil record was planted on earth by God to test our faith. But this would not account for the basic similarity of the chemistry of all life forms on Earth, unless one argued again that this was yet another test of our faith. The argument appears to get weaker each time it is used.

However, this chapter will only pick out from Darwin's legacy the salient parts of his theory that apply to eugenics.

Charles Darwin (1809–82) aged about forty-five.
(Mary Evans Picture Library)

The success of any animal species depends on various factors. Population numbers will depend on how well they breed together, how well adapted they are to their environment, and how adequately their food supply is maintained. Their numbers will not go on increasing indefinitely, because there will be opposing forces tending to decrease their numbers. There will be predators who consider them as food, there will be other species competing for the same food supply. Changes in the environment may increase the numbers of their predators, or a new disease may reduce their own numbers. So the population size will come into balance with these forces fluctuating first one way and then another, depending on circumstances. Selection by these natural circumstances will tend to favour the survival of the fittest to go on to breed and keep the population numbers high. The key factors in the equation according to Darwin are: being well adapted to the current environment, succeeding in the struggle for existence, allowing the fittest to survive in their habitat, and then the interbreeding of these fittest members to provide for the next generation. 'Fitness' in the evolutionary sense has no absolute meaning, but is only defined in relation to a particular environment.

But does natural selection actually work in practice? Consider the common brown rat, sometimes called the Norway rat (*Rattus norvegicus*). As a species it is highly successful, being found wherever man lives. They are able to breed every three to four months and can produce up to four litters per year, each containing between six and twenty-two young. They obtain a regular and constant food supply by living in close proximity to man. It has been frequently estimated that the rat population of the USA is approximately equal to the human population.

Rats will eat almost anything that human beings eat. Most serious is the damage they do to seeds and grain crops both before and after harvesting. Grain stored on farms is often not only eaten by rats but also rendered unsuitable for human consumption by being mixed with rat droppings. Food that reaches warehouses in cities is also eaten by rats and here the excess of damage over the amount actually consumed by the animal is even greater. Rats, although rodents, will also eat animal meat and have been known to kill several hundred baby chickens in a single night. In overcrowded urban conditions, such as slum dwellings, rats are prone to bite sleeping humans, especially children, and women dare not leave their babies alone in slum houses even for a few minutes for fear of a fatal attack by rats. Rats are also involved in the transmission of more than twenty human diseases including plague, murine typhus and Rocky Mountain Spotted Fever.

In view of the major economic and medical importance of rats, man has doctored their habitat by putting down poisons in close proximity to food storage places. By far the most effective poison has been warfarin, a tasteless and odourless chemical that interferes with the mechanisms of blood clotting. Rats that eat warfarin develop congested lungs and bleeding from the nose, eventually stop feeding and die. As a result of warfarin exposure rat numbers in northern England declined drastically over a period of six years. Then a surprising thing happened. After about 1960 the rat population started to increase again with recurring damage to food products. In 1958 on a farm in west Scotland a rat was found that appeared to be resistant to warfarin. By 1962 other rats caught on northern farms were being tested for warfarin resistance in the laboratory. They were fed warfarin and showed astonishing resistance, even when they were injected with the poison. This demonstrated that the resistance was not due to failure of absorption of the poison from the gut.

The underlying reason for the change turned out to be a single gene mutation that rendered the poison harmless. This change in gene structure may have existed in a few members of the rat population before warfarin was introduced into their environment but probably had no survival value. When exposure to warfarin occurred those few rats were now the ones better adapted to their environment; they competed more successfully with the rest of their kind who were warfarin sensitive, and eventually replaced them by breeding more numerously. So a slight genetic variation conferring an advantage can lead to some members of a species gaining the upper hand in the struggle for survival. Natural selection leads to the replacement of the less successful members of the species. This is entirely in accord with Darwin's theories proposed almost one hundred years before this particular example.

Today most rats are resistant to warfarin and it accumulates in their bodies if the chemical is used as a bait. In turn, this accumulated warfarin is poisoning the barn owl, the rat's natural predator. This is one reason why breeding pairs of owls have fallen from about 12,000 in the 1930s to only 4,000 pairs recorded in 1999 in England and Wales.

Two questions are raised by these studies: (1) are contemporary humans subject to the forces of natural selection?; and (2) do evolutionary forces operate in human societies? If so the weak, the inefficient and the sickly are destined to go to the wall. Failing to find breeding partners in competition with their own healthier kind, they should make no contribution to the next generation. Nature 'red in tooth and claw' is seen to be quite at variance with the Christian ethic of 'do as you would be done by'. It seems that by natural selection the meek and gentle of spirit are not going to inherit the earth. They will be eliminated in the struggle for survival. However, human beings are also exposed to a form of social selection in our choice of reproductive partners which can take into account such psychological factors as sociability, altruism and compassion. This may counteract to some extent the creed of living by 'tooth and claw'.

Two leading American industrialists have also commented on this issue in their time. J. D. Rockefeller said that the growth of a large business is merely the survival of the fittest; Andrew Carnegie claimed that struggle is inevitable and it is only a question of the weak going to the wall. This is perhaps more a social rather than a natural selection. Undoubtedly human populations are exposed to the same forces of natural selection as other animal populations. An environmental catastrophe, such as the collision of a large meteorite with the Earth, could change our environment so drastically as to wipe out our species. But generally speaking our psycho-social development has given us much greater control over our environment, so that chance fluctuations here will affect our population numbers much less than in other species. However, one historical example where one can see the effects of a type of directional selection in an early pre-Christian society favouring the survival of the fittest is the development of the Greek city-state of Sparta as described in Chapter 1.

Many of Darwin's contemporaries were concerned about applying evolutionary principles to human society. The British churchman Samuel Wilberforce asked in a debate with Thomas Henry Huxley (1825–95) 'whether the ape from which he is descended was on his grandmother's or his grandfather's side of the family'. To which Huxley replied: 'A man has no reason to be ashamed of having an ape for his grandfather. If there were an ancestor whom I should feel shame in recalling it would rather be a man . . . who plunges into scientific questions with which he has no real acquaintance, only to obscure them by an aimless rhetoric and distract the attention of his hearers from the real point at issue by eloquent digressions and skilled appeals to religious prejudice.'

T. H. Huxley (the grandfather of Aldous Huxley) was a fervent proponent and populariser of Darwin's theory on evolution but even he balked at the idea of applying it to the development of human societies. His argument depended on how one defines 'fittest' – there are so many moral overtones to the word. Fittest for what exactly? Darwin's theory implies fitness to provide the best offspring for the next generation – that is eugenics. Fitness for military conquest did indeed lead Sparta to a certain degree of success but at the expense of so many other things: poetry, drama, sculpture, literature, and all the other activities that make for an interesting life.

Huxley believed that a diversity of abilities and interests would make for a healthier society. In place of ruthless self-assertion and dominance by one individual or clique he advocated self-restraint; and in place of eliminating all competitors he suggested that individuals should respect and encourage their rivals since they bring to the work in hand a slightly different mix of qualities and abilities from their own. Reciprocal altruism should replace Darwin's gladiatorial theory of existence.

On the other hand Francis Galton, who was a cousin of Charles Darwin through his mother's side of the family, took a different view, as discussed in the next section.

(II) Galton: Darwin's Troublesome Cousin

'Diversity, controversy, tolerance' – in that
Citadel of learning we have a fort
that ought to armour us well.

– Marianne Moore, 'Blessed is the Man'

Francis Galton was something of a maverick character when young. He studied medicine at Birmingham (1838) but failed to qualify, and then tried again at King's College, London (1839–40). He left Birmingham General Hospital for a variety of health reasons: severe headaches and a 'very uncomfortable mind, but I shall soon get over all the hospital horrors'. In early 1840 he was studying medicine at King's, taking and enjoying courses in forensic medicine, chemistry, surgical operations and botany; but by July 1840 he suddenly announced to his mother that he intended to go to Liebig's laboratory in Giessen, Germany – 'Liebig is the first Organic Chemist in the world. In his Laboratory there is every opportunity for getting on, in addition to the certainty of a knowledge of German.'

He returned to England in October 1840 to enter Trinity College, Cambridge, to study mathematics. Again he had a bout of ill health: '. . . the reason why I write in pencil is as I am lying on my back I can't get a pen to write; I have been confined to my bed for some days, with rheumatism, not over-working, but will be shortly be released'. He claimed to have further bouts of recurrent ill health. In the May examinations at Trinity in 1841 he only obtained a third-class degree and by 1843 he suffered a complete breakdown in health. Unable to concentrate on maths due to severe dizziness and palpitations, he finally decided to give up reading for his maths Honours degree. Despite his many afflictions he lived to the ripe old age of eighty-nine.

Galton's early researches revealed the wide-ranging and disinterested intellectual curiosity which is both the lifeblood and hallmark of a great scientist. His research included developing

Francis Galton (1822–1911). (Mary Evans Picture Library)

the use of fingerprints to identify people, still employed today by the police. He worked on the inheritance of character traits such as intelligence, ability and sensory perception. He developed methods to study the inheritance of complex diseases such as tuberculosis, cancer and asthma. He made pioneering contributions to genetic statistics in which he came to analyse the question how far factor A in a parent contributes to factor B in the child. His reply was well ahead of the times: 'We must endeavour to find a quantitative measure for this degree of partial causation.'

Science should not concern itself with trifles, and Galton took on here some of the biggest problems of his and our own day. Replying to his own question of how to measure inheritance, in 1875 he surprisingly started to study the weights and diameters of sweet pea seeds between parents and offspring, quite independently of the famous Austrian geneticist Gregor Mendel (1822–84). From these data Galton constructed the first 'regression line' where the diameter of the parent seed is plotted against the diameter of the offspring seed in an attempt to measure how closely they are related. The correlation of the diameter of pea seeds between parent and offspring came out at about 0.33. This number means that the line of the graph slopes at about 33° to the horizontal. He arrived at the idea that the slope of the regression line would measure the intensity of the resemblance between parent and offspring. If there was no slope, the diameter of the offspring pea would be the same for all diameters of the parent pea and this would suggest no transmitted or inherited influence between the two. If the line sloped at an angle of 45° to the horizontal, the offspring's pea diameter would be exactly the same as the parent's pea diameter, supposing their average diameters were the same, and this would indicate direct inheritance of pea diameter. He also used this idea of correlation to study the inheritance of stature in 928 people correlated with the mean of their parents' heights, and found the regression slope to be 34° to the horizontal, suggesting some component of inheritance in determining body height. Finally, he observed that as the size of the parent pea increases, so does the size of the offspring pea, but the offspring is less likely to be a giant or a dwarf than the parent pea. This is Galton's discovery of the phenomenon of regression to the mean, which has been used by statisticians ever since.

The use of correlational analysis to attempt to define a functional relation between two variable quantities has been in use ever since Galton's day, and has been greatly refined into powerful tests by statisticians such as Pearson, Spearman, Kendall, Cramer and others.

It is a remarkable coincidence that both Galton and Mendel, both of whom were born in 1822, should have used sweet or edible peas for their studies of heredity. Galton said he chose sweet peas because he would not be troubled to the same extent by variation in size of peas within the same pod of the edible variety. It is not known whether he had heard of Mendel's work on the particulate inheritance of edible peas that was published in 1865 but went largely unnoticed in the European literature until 1900. Whether it is Galton's correlational calculus or Mendel's factorial analysis of legume heredity that is of greater importance for the future studies of genetics remains to be seen.

The major preoccupation of Galton's later working life was in the field of eugenics. As noted previously he invented and defined the term 'eugenics' as 'the science of improving inherited stock, not only by judicious and selective matings, but by all the influences which give more suitable strains a better chance'. He expressly wrote that there was no question of active suppression of the 'inferior' groups or classes of people as their decline would tend to happen naturally.

Galton strongly believed, following on from Darwin's theory of evolution by natural selection, that 'we should attempt to exert control over organic evolution in the same way as we exert control over the physical world and to direct it into channels of our own choosing'. This would provide practical applications for Darwin's theories and replace Darwin's idea of natural selection by a type of social selection.

The publication of Darwin's book *The Origin of Species* in 1859 and his theory of evolution by natural selection provoked bitter theological controversies with those who believed that God created man (and all the other animal species), as well as spirited debates on its application to social progress and reform. In Victorian England many social reforms were being put in place to help the poor, the weak and sickly members of society. These included the Anti-slavery Bill, Educational Acts, the institution of poor laws and workhouses, and plans to build new hospitals. 'Struggle for existence', 'survival of the fittest' and 'natural selection' were key evolutionary phrases used to attack the progress of such reforms.

Francis Galton was a fervent disciple of Darwin and his evolutionary theories. It even led him to adopt positions in which he really did not believe, such as Darwin's theory of pangenesis. According to this latter theory every cell in the body 'throws off gemmules or undeveloped atoms into the blood which are transmitted to the offspring of both sexes and can multiply by self-division'. These are Darwin's units of heredity. He knew nothing of Mendel's work. Pangenesis is a beautiful example of how a really great world-class scientist can be completely wrong. Scientific heroes, as is said of racehorses, should be 'tried high'. But it seems we cannot take our scientific heroes too seriously. We should never be dazzled by the big name. A more neutral attitude towards them is preferred nowadays. It is also an object lesson for the rest of us mediocre scientists – our ideas and pet theories can be hopelessly and disastrously wrong.

The intense admiration that Galton held for Darwin led him to an exaggerated respect for the authority of most of Darwin's opinions. He said in a speech to the Royal Society in 1886 that 'I rarely approached his [Darwin's] general presence without an almost overwhelming sense of devotion and reverence and I valued his encouragement and approbation more perhaps than that of the whole world besides. This is the simple outline of my scientific history.'

It is of great significance for eugenics that he valued Darwin's influence not so much for the insights it gave into the biological sciences but for the 'freedom Darwin gave us from theological bondage' (Linnean Society lecture, 1 July 1908). By inference he was more concerned with the gains for society and human progress that evolutionary theory might hold than for the advances of scientific knowledge of animal evolution. He believed that Darwin's theory of evolution gave us the ammunition to destroy the influence of men whose power depended on theology. Galton's eugenic ideas quickly became a labour of love, a sort of credo, despite the fact that there was no scientific evidence from any other studies to back them and there was no new technology to support them. The ideas from Darwin's writings on animal evolution appeared to be sufficient justification.

There were several areas, however, where Darwin and Galton disagreed. One was on the nature or mechanism of inheritance. Darwin firmly took the view of the blending nature of inheritance by use of his hypothetical gemmules. Galton appeared to hold two independent views. He agreed about the occurrence of blending inheritance for characteristics such as skin colour, quoting the example of children from mixed white and black marriages who show intermediate skin colours. But he also held the view of particulate inheritance for such characteristics as eye colour where children of light-eyed and dark-eyed parents tend more to take their own eye colours after one or other parent and rarely show intermediate or blended tints.

Conversely, Darwin was not entirely in agreement with all of Galton's theories but came round to some of them after much discussion. Thus, after reading Galton's book on *Hereditary Genius* published in 1869, Darwin writes to Galton: 'You have made a convert of an opponent in one sense, for I have always maintained that excepting fools, men did not differ much in intellect, only in zeal and hard work.' Darwin now appears to agree in principle that intellectual ability can be inherited. However, Darwin was more sceptical of Galton's ideas on eugenics. As one example, he writes to Galton in January 1873:

Though I see so much difficulty, the object seems a grand one and you have pointed out the sole feasible, yet I fear Utopian plan of procedure, in improving the human race. I will make one or two criticisms . . . the greatest difficulty I think would be in deciding who deserved to be on the [eugenics] register. How few are above mediocrity in health, strength, morals and intellect; and how difficult to judge on these latter heads. As far as I see within the same large superior family, only a few of the children would deserve to be on the register.

The value judgements as to who were more or less suitable to be on Galton's eugenics register for encouraging procreation, and what constitutes an improvement in inherited stock, were mainly left implicit in his definitions. He did draw up a tentative scale of how to estimate 'worth' corresponding to the virtues and values of Victorian England. Despite this inadequacy, Galton presented his methods for eugenics in great detail.

He proposed that family records should be kept to estimate the average quality of children, depending on their parents' occupation and ancestry. Extensive genealogical work should be undertaken on families who should be classified as: (a) gifted, (b) capable, (c) average, or (d) degenerate. This information should be stored in a State Eugenics Record Office. Early marriages for women of classes (a) and (b) should be encouraged and financial grants made available to them for producing numerous children. Such grants would be analogous with the award of grants for higher education for promising young adults or child allowances to couples in a particular category, both of which are practised today. In Vitrolles, France, in 1997 the Town Council introduced a law that only married couples of French origin would qualify for a newborn child allowance of about 5,000 FF (about $700). By 1998 this was judged to be illegal by the Administrative Tribunal of Marseilles on the grounds of racial discrimination. More recently (in 2001) Edmund Stoiber, the head of Germany's most conservative state, Bavaria, has mooted plans to sponsor German couples to produce more children. German parents would be paid DM1,000 ($484) a month for each child in the first three years of life. Needless to say this funding would not apply to children of immigrants. Similar measures were adopted in Singapore with a £1.2 billion package of incentives to be called the 'Baby Bonus'. Announced by their Prime Minister in August 2000, it is aimed at encouraging Singaporeans to have more babies; they will be rewarded for this by having £193 of government money placed into their joint savings account. The government will then match any further deposits made by the parents until the child's sixth birthday. If the parents have a third child the benefits will be doubled and the mothers will receive two months' fully paid maternity leave. The original ruling that this 'Baby Bonus' should only apply to university-educated Singaporeans was dropped as being too controversial. Despite all Galton's exhortation to others of high intelligence and abilities to produce as many children as possible, it should be noted that Galton himself had no children.

Galton further proposed that in competitive examinations for the Civil Service or other professional posts extra marks should be awarded for 'family merit' if the youth had a superior pedigree, as judged by the success of its members in their chosen profession. Rules of celibacy for Fellows at the older universities that existed before 1870 should be abolished, to allow reputedly intelligent academics to have children. Eugenic Certificates should become available for members of society and be kept in a Eugenic Records Office. A Eugenics Research Laboratory should be set up to gain more information on the inherited transmission of complex psychological traits such as intelligence, energy and perceptual abilities. Finally, state institutions should take some sort of action against the feeble-minded, habitual criminals and the insane. They should be segregated to prevent them from having children.

In the climate of Victorian England, Darwin eventually endorsed many of Galton's views, possibly seeing them as practical consequences of his own theory of evolution and natural selection. Although Galton's work is not quoted in *The Origin of Species* it is cited on more than ten occasions in *The Descent of Man*. Darwin came round to Galton's opinion that 'general intelligence, courage,

bad and good temper are certainly transmitted in families, and on
the other hand it is certain that insanity and deteriorated mental
power tends to be inherited'. He believed that 'natural selection is
affecting civilised nations, and the weak in body and mind are soon
eliminated'. He goes on:

> We civilised men, on the other hand, do our utmost to
> check the process of elimination; we build asylums for the
> imbecile, the maimed, and the sick; we institute poor-laws;
> and our medical men exert their utmost skill to save the life
> of every one to the last moment. Thus the weak members
> of civilised societies propagate their kind. No one who has
> attended to the breeding of domestic animals will doubt
> this must be highly injurious to the race of man. We must
> therefore bear the undoubtedly bad effects of the weak sur-
> viving and propagating their kind.

This is the great Charles Darwin writing in what would today be
seen as a very politically incorrect frame of mind.

But some contemporary geneticists do agree with Darwin that
natural selection is either not working or only operating ineffi-
ciently in human populations because of the improvements in
social and medical care, with a more uniform standard of living for
all. The use of welfare benefits to improve conditions for an economic underclass can lead them to produce more children than
other social classes. This in turn may lead to the survival and
expansion of the unfit (meaning the poor in this context) at the
expense of the fit. What effects this would have in the long term
for the future of the human species remain entirely conjectural,
depending on how one defines 'fitness'.

Darwin goes on to summarise his views in the concluding chapter of *The Descent of Man*:

> The advancement of the welfare of mankind is a most in-
> tricate problem; all ought to refrain from marriage who
> cannot avoid abject poverty for their children; for poverty is

not only a great evil, but tends to its own increase by leading to recklessness in marriage. On the other hand, as Mr
Galton has remarked, if the prudent avoid marriage, whilst
the reckless marry, the inferior members tend to supplant
the better members of society. Man, like every other
animal, has no doubt advanced to his present high condition through a struggle for existence consequent on his
rapid multiplication; and if he is to advance still higher, it is
to be feared that he must remain subject to severe struggle.
Otherwise he would sink into indolence, and the more
gifted men would not be more successful in the battle of
life than the less gifted.

The central tenet of eugenics seems obviously true: successful,
clever people will tend to produce good, clever children, and by
and large bad, stupid parents will tend to produce bad, stupid
children. And if the latter outnumber the former it will be a poor
lookout for society. Of course this depends on value judgements
on how to define success, goodness, cleverness, badness, etc. In a
eugenics primer printed in 1926 a pertinent question was asked:
'Does eugenics mean less sympathy for the unfortunate?' The
answer was: 'Eugenics does not mean less sympathy for the unfortunate; it does mean fewer unavoidable unfortunates with which to
divide your sympathy . . . this is true kindness, both to the victims
and to society.'

Francis Galton died at the age of eighty-nine in 1911, revered
and honoured to the last. He had his knighthood, he had been
elected as a Fellow of the Royal Society; and he had as he thought
discovered the new science of eugenics. A department and professorship with a eugenics laboratory had been set up for him at
University College, London. He had written sixteen books and
been awarded the Darwin Medal of the Royal Society in 1902 for
his 'numerous contributions to the exact study of heredity'. He was
also greatly pleased and flattered that most of his papers had been
translated into German, where the word 'eugenics' was added to
the German journal called *Rassenhygiene* (Race-Hygiene).

After the 1940s, however, Galton's fame was sadly on the wane. The scientific journal that originated in his department, the *Annals of Eugenics*, changed its name to the *Annals of Human Genetics*. His reputation as the founding father of genetics was totally eclipsed by that of Gregor Mendel. The departmental professorship at University College had its name changed to remove the word 'eugenics'. Galton became known more for introducing theories that led to the oppression of minority groups, to racism and to mass murder. Eugenics became a dirty word that has never fully recovered its original meaning. The atrocities committed in the name of eugenics are described in the next section but were never part of Galton's own plans for eugenics. The most he ever suggested was a strict segregation of criminals and the mentally retarded to prevent them from having children. To this day we separate men from women in prisons and mental hospitals partly for that reason and also to deprive them of things that normal society offers. We also use fingerprints for the identification of people and even more we use the statistic of cor-relation coefficients to analyse classes of data. All due to Francis Galton.

(III) The Nazis and Others

> This is a thing so evil that
> it could have come from nowhere save a mind
> capable of inventing the idea of evil.
> We are, professor, out of joint with the purpose
> of your work, whatever that point may be.
>
> – George Barker, 'Anno Domini'

Eugenic ideas were originated in Britain by Francis Galton and his followers. They captured the imagination of widely differing sections of the public, including socialists such as H. G. Wells, George Bernard Shaw and J. B. S. Haldane, as well as arch conservatives such as Leonard Darwin and Dean William Inge of St

Paul's Cathedral. But eugenic laws, such as those for the sterilisation of certain categories of people, were never enacted in England.

The range and diversity of opinions on the topic were enormous. To give some idea of the flavour of these: Leonard Darwin, son of Charles Darwin and President of the Eugenics Society, published a book on eugenics in 1928 advocating the use of sterilisation for the feeble-minded, habitual criminals and other people of 'inferior stock' so as to prevent the 'deterioration of our breed'. In the same year J. B. S. Haldane published a castigation of the sterilisation laws enacted in California in an article entitled 'Eugenics and Social Reform'. His advice was that if you want to check the increase of any section of society do not use sterilisation but either arrange to massacre them or to lavish an excess of wealth and freedom on them in which case they will probably destroy themselves. More seriously, as a biochemist and geneticist, he argued that compulsory sterilisation would be quite incapable of eliminating feeble-mindedness no matter how many generations were made infertile.

George Bernard Shaw, who had an appropriate word for everything, wrote enthusiastically about eugenics in his play *Man and Superman*, published in 1903. In it he dramatises the political need for a superman to be produced by 'good breeding' and an incessant struggle and aspiration for a higher existence. He devoted a chapter to eugenics in his *Everybody's Political What's What*. But it is very difficult to disentangle what he truly believed about the issues: his writing is so full of irony, satire, sarcasm and innuendo. What is one to make of his writing in 1944 that Hitler, by using his brains, had tried eugenic experiments very thoroughly in Germany? He continues: 'Let us hope that he [Hitler] will escape to enjoy a comfortable retirement in Ireland or some other neutral country.' Was Shaw being serious or ironic? I believe he was serious. After all Shaw was completely taken in by Stalin; he even celebrated his seventieth birthday in Moscow. He was captivated by what he saw of the Soviet Union. He praised Stalin and the regime to the skies. He never even suspected the reality of terror, oppression and famine in the Ukraine and the Volga lurking behind the prosperous facade.

In 1913 and 1934 Britain very nearly passed eugenic laws. Each time, however, compulsory sterilisation failed to reach the statute books. Many reasons have been given for this. There has been a stubborn torch of freedom burning in Britain's political thinking that has never been entirely quenched: 'Freedom is an English subject's sole prerogative' (Dryden, 1631–1700). Later in the nineteenth century, J. S. Mill (1806–73) published his book *On Liberty,* the bible of liberal democracy in Britain. It contains a magnificent chapter on 'Society and the Individual' in which he analyses the permissible limits to the authority of society over the individual. His finely constructed arguments lead to several major conclusions, one of which is 'that the only purpose for which power can be rightfully exercised over any member of a civilised community against his will is to prevent harm to others. His own good, either physical or moral, is not a sufficient warrant.' Mill gives many examples of 'illegitimate interference with the rightful liberty of the individual by society'. If the individual's action does not harm the interests or rights of anyone else, then he or she should not be accountable in any way to the state. The person may be advised, cajoled or persuaded to change his line of conduct but that is all – there should be no form of coercion. In Mill's view a major reason for 'restricting the interference of government is the great evil of adding unnecessarily to its power'. And in the long run 'the worth of the state is the worth of the individuals composing it'. 'A State which dominates its men in order that they may be more docile instruments in its hands even for supposedly beneficial purposes will find that with small men no great thing can really be accomplished.' To many people, those who imbibed these ideas with their mother's milk since the 1860s when *On Liberty* was published, the very idea of compulsory sterilisation by state laws must have been totally unacceptable, and the justification for such laws – an abstract principle to eliminate inferior stock, or prevent deterioration of our breed – a total anathema. Eugenics for such people was considered more as a matter for individual voluntary effort, and not a question for the state to direct the public as it thought fit.

Eugenics in America

The situation was different in North America. The USA began with the really grand inspiring statement in the American Declaration of Independence of 1776: 'We hold these truths to be self-evident, that all men are created equal, that they are endowed by their Creator with certain unalienable Rights, that among these are Life and Liberty' – but not happiness. It is there in the writing: that you have the right to pursue happiness, but it is implied that the chances are that you will never find it in the United States unless it be in the form of something like unlimited quantities of guns, computers, sex, fun or swimming pools.

Also, 'all men' did not seem to include African-Americans or other groups of immigrants. Admittedly Thomas Jefferson wrote that it was execrable that a statesman should 'permit one half of the citizens thus to trample on the rights of the other', and continued, 'whatever be their degree of talent [referring to the Negro] it is no measure of their rights. Because Sir Isaac Newton was superior in understanding he was not therefore lord of the person or property of others.'

As it developed American society became a mixture of differing ethnic groups and by the late nineteenth century restrictive laws were introduced to prevent free immigration of certain peoples. Beware of the alien, they sow the seeds of social disorder and rebellion. The first to suffer from this warning were the Asians. Two Chinese Exclusion Acts were introduced in 1882 and 1902. Other Immigration Acts soon followed to limit the influx of some European groups. There were many different reasons for these restrictions. Some were economic, arguing that free immigration would have an adverse effect on wages and that cheap labour might displace long-standing residents in their work. By 1900, in the coal mines of Pennsylvania, there was a multiethnic workforce of Polish, Italian and Irish immigrants working alongside each other with a high degree of mutual suspicion. Newer immigrants were considered as 'vermin' and were sometimes even stoned. It was believed that they would depress wages and make more profits for financiers such as Pierpont Morgan whose industrial conglomerate owned most of these mines. Other arguments put forward were social, that the immigrants would bring radical and unAmerican customs and values into society. And some were definitely racist. The status of one's own job would decline if one were to work alongside an immigrant or an African-American. What must depress the immigrant is the loneliness and uncertainty of a life with nothing fixed, and the status of a second-class citizen.

In 1921 the soon-to-be President of the USA, Calvin Coolidge, declared in a popular magazine his fear that 'biological laws show . . . that Nordic people deteriorate when mixed with other races'. European groups such as the Greeks, Russians, East Europeans and Jews were considered to be inferior to the earliest immigrants of Nordic and Anglo-Saxon stock. Discrimination against these newcomers and, of course, the resident African population whose ancestors had been slaves, led to measures to protect the resident white Americans from 'degeneration'. Strong legislation against African-Americans was introduced to produce apartheid in the southern states. Admittedly the Africans were no longer slaves but nor were they full citizens. In the early part of the twentieth century the legislature proceeded to pass laws declaring that some types of marriages, especially amongst the African-Americans, were void. So too were marriages between the feeble-minded, mental defectives and persons suffering from venereal disease. The laws then extended to sterilisation to prevent various classes of people deemed undesirable from having children.

The first sterilisation laws were passed in Indiana in 1907. By 1917 such laws had been enacted by fifteen more states, applicable to socially inadequate people, mental defectives and others. In Iowa the laws were particularly stringent. Here are some sections of a model sterilisation law of 1922 as drafted by H. Caughlin.

Sterilisation should be used for:

(a) A socially inadequate person who is one by his or her own efforts regardless of aetiology or prognosis, fails chronically in comparison with normal persons to maintain himself or herself as a useful member of the organised social life of the state; provided that the term socially inadequate shall not be applied to any person whose individual or social ineffectiveness is due to the normally expected exigencies of youth, old age, curable injuries, or temporary physical or mental illness.

   (b) The socially inadequate classes regardless of aetiology or prognosis are the following: (1) Feeble-minded; (2) Insane (including the psychopathic); (3) Criminals (including the delinquent and wayward); (4) Epileptic; (5) Inebriate (including drug habitués); (6) Diseased (including the tuberculous, the syphilitic, the leprous, and others with chronic infectious and legally segregatable disease); (7) the Blind (including those with seriously impaired vision); (8) the Deaf (including those with seriously impaired hearing); (9) Deformed (including the crippled); and (10) Dependants (including orphans, ne'er-do-wells, the homeless, tramps, and paupers).

Now you can see that goes rather far – especially when at the time contraception was doubtfully legal. Probably half the population of London could be legitimately sterilised if such laws were in force nowadays.

You may think that such laws would be impossible to pass today, but on 1 June 1995 the People's Republic of China passed a bill on 'Maternal and Infant Health Care'. It contains measures that hark back to America in the 1920s. These include 'termination of pregnancy if the foetus is suffering from a genetic disease of a serious nature or the foetus has any other defects of a serious nature'. A serious defect is defined as one totally or partially depriving the sufferer of the ability to lead an independent life. The couple must subsequently take long-term contraception or agree to undergo sterilisation. This could perhaps be seen as a practical way for China, with its limited health resources, to reduce the burden of disease in its population. With the one-child-for-one-family policy it is also important for the child of each family to be as healthy as possible. Further measures in the bill 'to avoid new births of inferior quality and heighten the standards of the whole population' include the deferring of marriage where genetic defects are likely to occur and compulsory sterilisation. 'China is in urgent need of adopting laws to put a stop to the prevalence of abnormal births.' Such ideas appear especially appealing in totalitarian states to provide a quick fix for difficult social problems.

State interference with the reproductive rights of individuals does occur to a slight extent in the UK. There is legislation that can deem parents unsuitable for social reasons to rear a child and therefore a panel of professionals, including social workers and doctors, will not grant a licence for the use of assisted reproductive techniques such as *in vitro* fertilisation.

In America in the 1920s opposition to these cruel and unnatural punishments began to develop. It came to a head in 1924 when a test case reached the United States Supreme Court, that of *Buck v Bell*. The case concerned three generations: a mother, Emma Buck, and her sister Doris her daughter, Carrie Buck, and a granddaughter, Vivian Buck, living in a 'Colony for Epileptics and Feeble Minded' in Virginia. They were all deemed to be mentally defective by the Binet IQ test, both mother and daughter scoring mental ages of about eight years old. It appeared to be a clear-cut case of inheritance over three generations. In fact Vivian, the granddaughter, was only seven months old at the time of 'testing' but was reported to have a 'look' about her which was not quite 'normal'. This was sufficient evidence to convince Justice Holmes, who declared that 'three generations of imbeciles are enough'. Doris and Carrie were duly sterilised. The *Buck v Bell* decision was a triumph for the eugenic movement in America and the American Constitution on this issue was now clarified. By 1931 twenty-seven other states enacted sterilisation laws to prevent the undesirable classes from reproducing. Most of these laws provided for the voluntary or compulsory sterilisation of certain classes of people thought to be insane, feeble-minded or epileptic. Some were extended to habitual criminals and moral perverts. In most states these extreme measures were not enforced, except in California where about 10,000 people had been sterilised by 1935; in all the other states some 20,000 people were sterilised. As a postscript to the case of *Buck v Bell*, Carrie's daughter, Vivian, went through second grade at high school, where her teachers reported her to be very bright. Doris Buck later married a plumber and tried to have children. She was never told the real nature of her operation, simply that it had been for a burst appendix. When she learned the truth she said: 'I broke down and cried. My husband and me wanted children desperately. We were crazy about them. I never knew what they'd done to me.' A heart-rending lone cry against the powers of the state.

An unfortunate side-effect of the extensive American eugenics programme was that it provided support for the Nazis and their 'eugenic' measures decreed ten years later.

Eugenics in Germany

The 'eugenic' movement in Germany was well under way before the Nazis came to power. 'Eugenics' here is placed in inverted commas and should be read as such for the rest of this section because its appropriation in this context is a perversion of the original meaning of the word. It is like saying that extermination and euthanasia are synonymous. Euthanasia means providing a painless way to die, voluntarily requested by the person, whereas extermination means mass murder.

In 1889 a group of stalwart Germans, possibly under the influence of Galton's ideas and organised by Nietzsche's sister Elizabeth, set sail for Paraguay to found a Teutonic colony. She was following the lead of her prominent husband, Bernhard Forster, who was both anti-Semitic and nationalistic, and together they wanted to found a pure Aryan republic. Remnants of the colony exist to this day. General Alfredo Stroessner, the son of a German immigrant, set up a military dictatorship in 1954. He imprisoned or killed his political opponents and dominated the nation's congress and law courts. Half of the entire national income was spent on military establishments to bolster and preserve its dictatorial authority.

Many of Galton's papers on eugenics were translated into German and published in the *Archiv für Rassen und Gesellschafts-Biologie* in 1906. By 1931, two years before Hitler came to power, the German Society of *Rassenhygiene* had added eugenics to its name. This society became associated with mystical concepts of the purity of the Nordic race and the fear of degeneration of Europeans in general and that of the German *Volk* in particular by alcoholism, syphilis and the overbreeding of the feeble-minded or people from the lower strata of society. General eugenic ideas linked to racism and nationalism soon became identified with the developing Nazi ideology. The word 'race' has since become such a slippery concept that it has never fully recovered from the pejorative effects of Nazi propaganda. Some scientists now refuse to use the word and adopt the term 'ethnicity' to describe the cluster of shared characteristics that are found in a particular group of people (such as dark skin in Afro-Caribbean groups, or the inner canthus skin folds of eyes in Asian groups).

German Fascism was emptier of theoretical principles than Russian Communism, but was frighteningly dogmatic about its racial policies. The perverted use of eugenics in Germany to fulfil these aims during the period 1935–44 appalled many contemporary evolutionary biologists such as J. S. Huxley, J. B. S. Haldane and Gaylord Simpson. This was a primary reason why Galton's work and writings fell into such disrepute and invited such opprobrium, although his ideas and proposals were much more moderate even than those of the Ancient Greeks, as discussed in Chapter 1. There may have been a certain amount of hypocrisy in this attack on Galton because both Haldane and Huxley were active members of a dictatorial colonial empire in which racial superiority was itself a strong theme. When Mahatma Gandhi was asked what he thought of Western civilisation no wonder he replied, 'I think it would be a good idea.'

Hitler did not justify his social policies on the basis of social Darwinism or eugenics. There is no need to dwell on Hitler's lack of education, but he had a deep and genuine ignorance of biology and genetics. No reference to such subjects can be found in his books, *Mein Kampf* or *Table-Talk*. His social ideas appeared to derive mainly from the nineteenth-century German philosophers Schopenhauer, Hegel and particularly Nietzsche, who is quoted several times in *Table-Talk*. Nietzsche's advice finds echoes in Hitler's writings: 'I counsel you not to work but to fight . . . not to make peace but to conquer. Let your work be battle. War and courage have done greater things than charity. Have the will to power . . . the will to be master.' Nietzsche praises warfare because it makes men more 'barbaric' and more 'natural'. He praises the supermen whose thinking is of value not for its truth but for its effectiveness; he scathingly attacks Christianity and derides the 'virtuous stupidity' of the majority.

The case is different with regard to the German biologists, anthropologists and geneticists of the period between 1933 and 1945. They actively invoked eugenic principles to justify the social policies of the Nazis. In 1935 measures were enforced to promote a healthy German stock. A marriage law was instituted to protect the heredity of the German folk and prohibited the union with 'alien races' and those affected with hereditary diseases. State Marriage Loans were awarded to persons with official Certificates of Health only obtainable through a Health Sanitary Board. These grants were increased by 25 per cent after each successful live birth, to promote larger German families.

Families with four or more children were classified into the following categories: (1) antisocial, (2) bearable, (3) average, and (4) high-quality. Privileges such as educational grants were only awarded to categories (3) and (4). The authority for these laws rested with the 'Council of Experts for Population and Race Policies' whose members included human geneticists and anthropologists (Prof. Dr E. Rüdin, Dr Med. A. Ploetz, and Prof. Dr F.

Lenz); statisticians; lawyers; and top-ranking politicians, the most prominent being Heinrich Himmler, Reichsführer SS. Their ultimate aim was to possess a family file on every individual in the Third Reich. A 'Reichs-Committee for Recording Severe Hereditary and Congenital Diseases' was established in 1939 whose role was to register all new births with malformation and to consider 'euthanasia'. The first meeting of a 'Euthanasia Commission' occurred in 1939 to discuss the means of mass extermination of hospital in-patients and other 'useless feeders'.

In practice a system of courts was established with a tribunal composed of a judge, a medical officer and a doctor with specialist training in racial hygiene. The medical officer could initiate proceedings as well as adjudicate cases, and doctors were in the majority. Heinrich Gross was one such medical officer in charge of the children's ward at Spiegelgrund in Austria from 1941 to 1944. This was one of the thirty to forty 'euthanasia' clinics for children set up by the Third Reich to carry out the 'eradication of the pathological genotype'. Witnesses said that children there were exposed on balconies during winter nights and given injections and sedatives to lower their resistance to disease. After the war, all but one of Gross's colleagues were convicted of war crimes and the head of the clinic, Ernst Illing, was hanged.

Other laws were introduced against habitual criminals, homosexuals and antisocial elements in the society. Penalties included compulsory sterilisation, compulsory abortion and detention in work camps.

Thereafter the situation deteriorated rapidly, taking so-called eugenic views to the most perverted and macabre of conclusions. Registration of antisocial families, cripples, Jews, gypsies, vagrants, homosexuals and socially inferior individuals was enforced, with a view to their mass extermination. Following various eugenic laws against women, approximately 200,000 females were compulsorily sterilised. Finally, to protect the German blood and honour (*Gesetz zum Schutze des Deutschen Blutes*), measures were put in place for the total extermination of 'inferior races'.

By a gradual and insidious series of compromises many

geneticists and scientists, perhaps to stay in post, to curry favour with the powers-that-be, and to preserve their pension benefits, modified their opinions to coincide with the lunatic wishes of Hitler. One often-quoted case is that of Professors Hallervorden and Spatz who were directors at the prestigious Kaiser Wilhelm Institute of Brain Research in Berlin. In 1922 they had described a curious form of familial brain degeneration, often called to this day 'Hallervorden–Spatz' disease. Their further work required a supply of brains for microscopic examination; what better place to acquire them than from the Nazi extermination centres? Hallervorden visited these and provided technical help from his department for removal of victims' brains. No protest, no indignation, no outrage, only the following:

> I went up to them: 'Look here now, boys, if you are going to kill all these people at least take the brains out, so the material could be utilised.' They asked me: 'How many can you examine?' And so I told them an unlimited number – the more the better. I gave them fixatives, jars and boxes, and instructions for removing and fixing the brains and they came bringing them like the delivery van from a furniture company. There was wonderful material among those brains, beautiful mental defectives, malformations and early infantile diseases.*

Both Spatz and Hallervorden were reinstated in their posts at the Institute in 1948 after the war and continued 'honourably' until their retirement, dying in 1965 and 1969 respectively.

The further consequences of the Nazi policies have been extensively written about elsewhere (see 'The Nuremberg Doctors', *British Medical Journal*, 1996). It has become our recurring historical nightmare. Suffice it to say that more than twelve million people belonging to 'inferior races' died in mass exterminations.

* Quoted by Muller-Hill as Hallervorden's response to interrogation by American officers at the end of the war.

The doctrine of forgiveness of sins is one of the more difficult aspects of Christianity, the issue always being how far to go. Should forgiveness ultimately preside over all our human acts? Other religions, such as Buddhism, see it differently – it is not a question of expiating your sins but of developing techniques either to eliminate or at least to control them. If all our sins are to be forgiven, society would become impossible. Gluttony, sloth, concupiscence – yes, perhaps forgive. Extreme cruelty, enslavement, torture, rape and mass murder by the Nazis demanded some form of retribution at the end of the war. The trials at Nuremberg in 1945–6 attempted to lay the burden of guilt on the perpetrators of these crimes against humanity. Most of the defendants, including Hermann Goering, who was one of Hitler's designated successors, pleaded not guilty either on the basis of ignorance of what had happened, or because they were simply following their duty of absolute obedience to the orders of the Führer. Goering excused his participation in the founding of the concentration camps that led to the extermination camps in these terms: 'When the need for creating order, first of all, and removing the most dangerous elements of disorder directed against us now became evident, I reached the decision to have the Communist functionaries and leaders arrested all at once . . . The Nuremberg Laws were intended to bring about a clear separation of races and in particular to do away with the concept of a person of mixed blood in the future.' Sentenced to death by hanging, he committed suicide in his prison cell by taking a cyanide pill before going to public execution.

Eugenics in other countries

The Nazis were not alone in considering the Nordic as the ideal biological type. Other north European countries including Denmark, Finland, Norway and Sweden started sterilisation programmes of their own from 1926 onwards. Even more remarkable is that some of these laws remained on the statute books until 1976. In Sweden between 1935 and 1976 no fewer than 60,000 young women were compulsorily sterilised. They included mental defectives, or those otherwise handicapped and incapable of looking after their children. People having undesirable racial characteristics or otherwise inferior qualities such as very poor eyesight or an 'unhealthy sexual appetite' were also sterilised. The major reason for doing this was a benevolent form of social Darwinism to prevent 'degeneration of the race'. It would stop the feeble-minded from breeding in greater numbers than energetic people of superior stock. It would also save the state the heavy costs of welfare benefits. Our poorly endowed neighbour was to be sacrificed for making the mistake of being a lesser person than us.

In Sweden, many dreadful situations occurred. One young girl was sent to a school for the educationally subnormal and whilst there was sterilised. She shouldered her burden silently. Later she was found to be of normal intelligence but it was revealed that her parents had been too poor to afford to buy her glasses as a schoolgirl and her eyesight was so bad she could not see the blackboard properly. For that she was labelled as mentally defective. Just as bad were teenagers as young as fifteen who were sterilised, some without their parents' consent, for inadequacies as vague as 'lacking judgement' or having 'no obvious concepts of ethics'. Children in special educational schools and reformatories were induced to have the operation as a condition for their release. Pregnant women seeking an abortion because of a deformed foetus were also told to consent to sterilisation.

Such women have been haunted by this treatment ever since and it has been a continuous source of shame for them. Some are now suing their government and obtaining small amounts of compensation of up to £13,000. It was reported as recently as two years ago that the Swedish Social Affairs Minister, Margot Wallstrom, refused an application for compensation from one victim on the grounds that the policy had been legal at the time. Is it enough just to help these victims up with a single financial award? Do they not need more continuity of support and help afterwards?

Why did such laws remain on the statute books in Sweden for

so long? The horrific example of the Nazi eugenic programme apparently did not deter the Swedish legislators or their advisers. Much of the failure to react has been put down to inertia but the exploitation and victimisation of a minority group clearly appeals strongly to certain political minds.

White Australians were not immune from such ideas of state coercion. Apart from massive ethnic cleansing of sections of the indigenous (aboriginal) population, the New South Wales Legislative Assembly in 1927 advocated laws to segregate 'mentally defective persons for the purpose of effecting an improvement in the race stock'. 'It has been estimated that in one generation mental deficiency would be reduced by half if sterilisation were legalised. In Australia there is the power to make the race we wish. It should not only be a white race, but a race of the best whites.'

A Racial Hygiene Association was formed in Australia in the 1920s that advocated the compulsory exchange of health certificates between couples who intended to marry. Women were reminded 'that motherhood was "not an instinct but a science"'. From the 1920s to the mid-1960s most of Australia's six states pursued a racist policy of separating aboriginal children from their families. As many as 55,000 children may have been forcibly removed and put into foster homes, many run by the churches. Two reasons were given for separating such children from their families: 'breeding out the colour' was how some officials at the time described the aims of the policy. Alternatively they indicated that aboriginal parents wanted 'to give their children a Western education'. The reader perhaps can best decide which is the most likely.

Sterilisation of handicapped women is still reported from time to time in the tabloids today. Fourteen mentally handicapped young women aged between twenty-five and thirty-two were sterilised without their consent or agreement of their families between 1991 and 1998 in the town of Sens, 120 kilometres south of Paris. The Association for the Defence of the Handicapped in the Yonne district is demanding a criminal investigation of this because, according to law, performing such operations without consent constitutes a form of assault and mutilation. It is punishable by up to thirty years' imprisonment.

More serious were the reports of secret sterilisation experiments in Norwegian hospitals involving mentally retarded patients during a twenty-year period up to 1994. The aim was to assess the effects of irradiation on different parts of the body, including the sex organs. The possibility of a nuclear bomb attack on the West led to the need for greater understanding of the health effects that nuclear irradiation might produce. Fredrik Mellbye, an eighty-one-year-old colleague of the former Director of Norway's Health Services, revealed the details of the project to the Norwegian newspaper *Dagbladet*. 'I cannot remember that anyone at any time put their foot down to stop what was happening,' he said. 'Both authorities in the health services, psychiatrists, and other doctors knew what was going on.'

The point of raking over these ghastly events is not simply in the spirit of 'lest we forget' but more for the practical purpose of early recognition of the symptoms of a recrudescence of this sort of political activity. The fight against intolerance, authoritarianism and the mentality of men who generated these past events is never ending. We think of our own generation as standing at the apex of civilisation from which the deficiencies of previous ages may be condescendingly viewed in the light of our own scientific progress. In particular the latter half of the twentieth century in Europe, with all its many Declarations of Human Rights, was going to improve on all others. It hasn't. Certain problems were never going to occur again: terrorism, torture, indiscriminate killing. They have. Within the past five years the horrors of bombing of crowded cities (Belgrade), of ethnic cleansing and photographs resembling the Nazi concentration camps, have again come to light in, of all places, Europe – from the former Yugoslavia. The fact that it is happening in the post Nuremberg era and in Europe is almost inconceivable. When a group of individuals, politicians, sociologists, scientists, or whatever, start making dogmatic assertions and recommending punitive actions against a defenceless minority group, basing their opinions on pseudo-genetic jargon, then even a blind person can see that important values are at stake. A clique of like-minded men can gather around them to become the bane and ruin of civilised values and behaviour.

6

The Individual and the New Eugenics

Underneath all, individuals,
I swear nothing is good to me now that ignores individuals,
The American compact is altogether with individuals,
The only government is that which makes minute of individuals,
The whole theory of the universe is directed unerringly
to one single individual – namely to You.

—Walt Whitman, 'By Blue Ontario's Shore'

Privacy is becoming a hot topic. Your VDU flashes out its dry commands to insert text, to point and click, followed by encrypted reports flying around the Internet. Will meddling fingers start to record your deciphered credit card digits used to make your last purchase on the Internet? Or will your on-line banking codes be hijacked to lead to fraudulent purchases from your bank account? If so you only stand to lose money from your bank balance. For loss of confidentiality of your personal medical records or details of your genetic code you stand to lose much more. The job you applied for; the mortgage you want for buying that house; your rights to drive a car; the work you already do; even your personal freedom – all these can be put at risk if your medical details fall into the wrong hands. So quite strict rules governing medical confidentiality have been established.

Individual confidentiality and privacy

At present genetic information is collected for two major reasons – medical and legal. For medical uses there are ethical and professional guidelines laid down by the General Medical Council (GMC) in the UK. These stress the obligation for confidentiality, but recognise that there can be disclosure if it is in the public interest. Disclosure can be ordered by the courts, and may be permitted if required by statutory authority. Disclosure in other exceptional circumstances is recognised. These include issues relating to public interest when there is a risk of death or serious harm to the patient or another person. The General Medical Council makes further exceptions for breaking the rules of privacy for unfit drivers, and for patients who by themselves are unable to consent to treatment because, for example, of learning difficulties. This is a form of professional regulation that seems to work well.

Although there is no specific legislation governing medical confidentiality in the UK, the courts have dealt with the issue on a number of occasions. In the case of *W v Egdell* the public interest was held to justify the release of a medical report on a prisoner to a parole board by the doctor who had made the report. The case concerned W, who was a violent criminal on trial for murder but who wanted to be released on parole. His solicitor asked Doctor Egdell to prepare a report on his mental state. The report was unfavourable and was not used by the solicitor. But Egdell was sufficiently worried about W's mental state to send the report to the Secretary of State, who passed it on to the parole tribunal. W, on hearing about this, sued for breach of confidence. The court held that there was indeed a duty of maintaining privacy, but that in this case an exception could be made in the interests of the safety of the public. This case is important because it helps to clarify where the balance between maintaining privacy and risk of harm to the public should lie.

Disclosure is limited to those who can protect the public and only when there is a real risk of danger to the public. For example, in 1999 and 2000 there were requests for public disclosure of the medical reports on former Chilean dictator General Augusto Pinochet by human rights agencies including Amnesty International. This was before attempts were made to extradite him from the UK to Spain to face charges of crimes against humanity. The Secretary of State Jack Straw decided that there was no risk of harm to the public posed by these reports and their privacy should be maintained. He did not even use the word 'senility' in his report to Parliament because this could have been considered a breach of the rules of medical confidentiality. Subsequently a court ruling under Lord Justice Simon Brown maintained that the Home Secretary should disclose the medical records to the four states requesting extradition (namely France, Spain, Belgium and Switzerland) but under conditions of strict confidentiality. This decision was considered as a significant victory for the countries and human rights groups who wanted Pinochet to face charges of perpetrating torture against Spanish and other citizens during his 1973–90 regime in Chile.

In general, medical reports may not be disclosed to third parties. Court decisions confirm the seriousness of the obligation for medical confidentiality, but also that it is not absolute in exceptional cases.

Various international treaties ranging from the Universal Declaration of Human Rights in 1948 by the United Nations to the Convention for the Protection of Individuals with regard to the Automatic Processing of Personal Data (1981) may provide general principles of guidance for the use of medical and legal databases.

Legal databases are currently being built in much the same way that fingerprints are collected as a means of identification by forensic scientists. The aim in Britain is to register the entire criminal population on a DNA database by 2003. These are for crimes ranging from car theft, to use of drugs, to more serious crimes such as murder. So far about 940,000 samples from criminals and about 83,000 samples from the scenes of crime are being held. Such material was only to be stored permanently if the suspect was subsequently convicted; the rest should be destroyed. A team of Home Office inspectors has, however, found that the database is holding profiles of thousands of those no longer under suspicion. The database is now officially being extended to contain DNA samples from people who have never been charged or convicted of any criminal offence. This would increase the database size from about one million to more than three million in three years.

Civil liberty groups are alarmed at the number of DNA samples that are being collected and the future use that might be made of them. Unlike fingerprints, an individual's DNA can be put to many other uses than simply the identification of a criminal at the scene of the crime. Any gene of interest can be amplified and commercially exploited, as described on page 114, or the whole genome can be scanned for all personal biological details coded by the genes.

Most other jurisdictions do adhere to the principle that genetic details should not be kept of those individuals not charged with an offence or not convicted. Further questions related to this are: what other information should be stored in the database, and who should have access to it? At present only demographic details and genetic markers are being used, but there is the potential for recovering the complete information on all the genes that make up that individual.

Another government database in the UK is being built to collect all the details of donors and recipients of sperm or eggs used in assisted reproductive techniques that currently require a licence from the Human Fertilisation and Embryology Authority.

International declarations do provide a useful set of guidelines on the use of DNA databases, but they also contain many errors and ambiguities. They focus on the protection of human rights rather than remedies for injuries subsequently suffered. For example, in 1996 UNESCO published a Declaration on the Human Genome and Human Rights in which it is stated that the human genome is the 'common heritage of humanity'. This was presumably intended to rule out the possibility of taking out patents on parts of the human genome. But the statement can be taken to mean that the genes of any individual member of the species *Homo sapiens* belong to all humanity. Or could it mean that all humanity may make use of any individual's genetic code? In either case there is no obvious provision for maintaining confidentiality of a person's genetic make-up.

Domestic law or professional regulation will usually be applied because of the ambiguities of the various European declarations, but interaction with domestic law is such that condemnation of the law in a particular jurisdiction by the European Court of Human Rights at Strasbourg will often lead to a change in domestic law. A good example is the case of *Gaskin v UK* where the right to know one's identity as laid down by the Court led to the enactment of the Access to Personal Files Act 1987. This case concerned an applicant placed for adoption who spent an unsettled childhood being moved from one temporary home to another, eventually ending up in prison. In an attempt to understand what had gone wrong with his life the applicant asked to see his local authority's files. This was denied on the grounds of confidentiality of the reports made by the social workers and foster parents, who did not anticipate that the applicant would ask to see their comments. However, the European Court at Strasbourg criticised the UK government for failing to provide objective mechanisms for the assessment of the reports in order to establish rights of access to information. In the course of judgement the Court enunciated the principle that every individual has the right of access to information relating to his or her own personal life. The European Convention on Human Rights has now become part of UK law since the enactment of the Human Rights Act of 1998.

Genetic knowledge is family knowledge. This raises particular problems where screening or the identification of genetic disease in one individual has implications for other family members. To take an extreme case: if one of an identical twin pair finds that he is a carrier for a serious disease gene, does the other twin have a right to know about this or is the information strictly confidential to the first twin? The rules of doctor and patient confidentiality require that the patient's wishes are paramount. Although there are no specific legal rules giving exceptions to this in the interests of the health or welfare of other members of the patient's family, there is no shortage of professional recommendations. Unfortunately the advice varies according to the source it comes from. There is general agreement that it is appropriate to persuade the patient to give consent to inform others who may become affected by diseases such as bowel cancer or Huntington's dementia. If the patient agrees, the doctor may then inform the relatives.

However, if there is refusal to cooperate in informing those family members who may be affected, the expert advice is contradictory. The House of Commons Science and Technology Committee report on human genetics (1995) gives precedence to the doctor's duty of confidentiality to the patient rather than to all those who may share a common genetic inheritance, stating that 'the individual's decision to withhold information should be paramount'. The Nuffield Council on Bioethics, however, concludes that in exceptional circumstances 'health professionals might be justified in disclosing genetic information to other family members, despite an individual's desire for confidentiality'. The British Medical Association advises its members to respect the guidelines on confidentiality laid down by the General Medical Council. This allows for disclosure of information without consent, where 'necessary in the public interest where a failure to disclose information may expose the patient or others, to risk of death or serious harm'. It has been claimed that maintaining confidentiality in genetic matters will have no adverse effects on the outcome for other members of the family. But this is not so. It may affect the relative's ease of access to regular monitoring and the use of early therapeutic measures for such serious diseases as cancer, heart disease and the dementias.

There are other areas where invasion of genetic privacy could be applied unfairly. For example, in adopting children prospective couples have been screened for genetic disease that might impair their ability to act as parents. In one reported case a couple were denied adoption because the wife was 'at risk' of developing Huntington's disease. She was considered unsuitable to provide long-term care for the child.

It will be difficult to legislate on the issue of medical confidentiality in genetics until a consensus develops amongst health professionals, politicians and the public. The issue of whether information about genetics is family information to be shared amongst those likely to be affected, or whether it is personal information relating to a particular patient, remains open. The point is, surely, that this information is both of these. As such it is arguable that a duty towards all those likely to be affected must be recognised, even when they are not patients of the doctor who has the information. Such a duty has been recognised in Californian law in the case of *Tarasoff v Regents of the University of California*. Here a psychiatrist, Dr Lawrence Moore, was held to owe a duty of care to a young girl, Tatiana Tarasoff, who was not one of his patients. She was eventually murdered by one of his patients, a Mr P. Poddar. The patient had previously revealed extensive murder fantasies during psychotherapy but Dr Moore did not take any action. However, legal opinion is that such a duty will not be held to exist by the English courts in the absence of legislation.

Control and property rights over DNA samples

Although it is claimed that genetic testing is no different from other biochemical assays (such as the estimation of blood glucose or fats) it differs in three major respects. First, the results of the test can be of direct importance for first-degree relatives. Secondly, the extracted gene sequences can be cloned and used for purposes other than those originally intended. Thirdly, they can become of commercial value. For these and other reasons France has passed quite stringent laws regulating the use of DNA for the study of genetics in populations. These laws are: informed consent must be obtained for participating subjects, approval of a local ethical committee must be obtained, and a nominated principal investigator who must be a medical practitioner to be responsible for the study. These three points are generally accepted in the UK although they are not binding by law. French Law (of July 1994) has gone further and requires an accreditation of the place where the work is to be carried out, a declaration of persons participating in the work to be made to a national registry, and regulation of the use made of the extracted DNA other than for its declared use in the original research plan. The national or international exchange of extracted DNA is restricted and has to be requested from a specialist committee at the National Institute for Health and Medical Research (INSERM). This cumbersome system affirms many of the declared principles of confidentiality, non-commercial exploitation, and limitation of the use of genetic testing to specific research proposals. But it becomes restrictive and inhibitory for the organisation of large-scale studies on the genetics of large populations.

As genetic markers become more precise, the original gene described in the research plan may now become of less value, but to study a new gene would require submitting a new research proposal. Also the DNA collected for one set of tests may be of use for analysis of another set of markers, even possibly for a completely different project. Again this would require submission of new and separate research proposals. The simple exchange of samples between laboratories for quality control of assays is prohibited unless permission has been obtained from a national committee. By contrast in the UK there is no legislation so far directly related to the use of DNA in population screening but there is a code of practice which, if not followed, could lead to a challenge from such bodies as the General Medical Council or the British Medical Association.

In 1998 the Ethics Committee of the international Human Genome Organisation (HUGO for short) recommended that, provided informed consent has been obtained and conditions for privacy and confidentiality in the collection, storage and use of the DNA are in place, the extension of its use to other projects would be considered ethical. Special consideration should be made for access to the information by immediate relatives where there is a high risk of transmitting a serious disorder. And unless authorised by law there should be no disclosure to institutional third parties such as employment agencies or insurance companies without appropriate consent.

Who owns the genome?

With regard to the property rights of the DNA sample, in UK common law there is no personal ownership of the human body or its parts. There are only two basic types of benefit that we have in our lives: the possession of our own bodies and minds, and the enjoyment of our objects and properties. The intricate and elaborate rules and laws governing the protection of our property – our homes, our money and our possessions – far exceed those that relate to governing our own bodies.

Most people would not doubt that they own their own bodies, especially their genetic inheritance. But this is not so. In common law you do not have ownership of any parts of your body while you are alive – although of course you do have certain personal rights. And after death no one owns the cadaver although our next of kin have an obligation to bury or cremate us. Quite why the common law has taken this position is not clear, but it is reflected in case law throughout most common-law countries such as the USA, Canada, Australia and New Zealand. John Locke (1632–1704), whose writings laid the foundation for the American Declaration of Independence and the American Constitution, stated in 1690 'that every man has property in his own person. This nobody has a right to but himself.' It is the common sense point of view. But the rules seem to have been changed in the eighteenth century. Perhaps they were then reinforced as a result of the abolition of slavery in the nineteenth century when it was made illegal for one person to have ownership over another. This is perhaps why people feel more comfortable about allowing interference with the genes of someone else rather than the purloining of another person's property.

That there are no ownership rights in English Law over your body parts was confirmed in 1990 in a court case about the ownership of body parts, corpses and genetic material. However, lack of property rights must be distinguished from personal rights, which do exist. This means that a legal action based on rights for the use, without consent, of genetic material by someone else will succeed. The point is that consent is required both ethically and legally.

The leading case on commercial exploitation of body parts such as genetic material was that of *Moore v Regents of the University of California* (1990). An Alaskan businessman, John Moore, found that his own body parts had been patented without his consent or knowledge by the University of California at Los Angeles (UCLA) and licensed out to the Sandoz Pharmaceutical Corporation. This came about because he was suffering from a rare cancer and the doctors at UCLA found that his spleen was producing a protein that facilitates the growth of white blood cells and could be a valuable anti-cancer agent. The university grew cells in cultures from Moore's spleen and patented their 'invention' in 1985. It was estimated that these cells could be worth about $3 billion. Moore subsequently sued the University of California, claiming property rights over his own tissues. The California Supreme Court ruled against Moore, maintaining that he had no such rights since body parts could not be bartered freely as a commodity in the marketplace. The Court did say that the 'inventors' had a responsibility to inform Moore of the commercial potential of his tissue and for that reason had breached their position of trust and might be liable for damages. But in the final analysis the Court upheld the primary claim of the university that the cells could be justifiably claimed as being the property of the University of California and not of Mr Moore.

This type of result has encouraged other institutions to exploit the value of various body parts. It has led to seemingly ridiculous patents being awarded to biotechnology companies. In 1999 the American company Celera filed provisional patents for no fewer than 6,500 genes and their fragments. Another company, Incyte, aims to patent more than a million fragments of genes (called expression sequence tags) without any clear idea what they are going to do with them but just so as to have control of the resource. Yet another American company, Biocyte, was awarded the ownership of all human blood cells coming from the umbilical cord of placentas for use as therapeutic agents. They were awarded the patent because they had invented a method to isolate and deep freeze the cells. The patent is so broad that it allows this one company to refuse the commercial use of any umbilical blood cells by anyone not prepared to pay the patent fee. Another broad patent was awarded to the American company Sytemix by the US Patents and Trademark Office for the therapeutic use of all human bone marrow cells without the company having 'invented' anything to alter or to engineer the cells. The UK office granted patents with rights lasting for seventeen years to the Roslin Institute, Scotland, to cover some aspects of the use of cloning techniques employed to create Dolly the lamb. These also included claims to cover aspects of therapeutic cloning of human cells. Some enterprising company of the future will no doubt patent all the cell lines derived from the gonads and so gain total control over the reproductive capacity of the human race. A patent fee could then be liable for every newborn baby.

Some people have moral objections to the whole idea of patenting 'life'. It is like patenting water or air. By all means patent inventions which have needed ingenuity to develop and are both novel and useful. If they lead to successful industrial applications the patent provides financial rewards to the inventor and repays his time, efforts and possibly the venture capital needed to get the research started. Patents last up to twenty years and provide protection for the inventor's intellectual property. Other people are stimulated to find better ways of doing things and improve the methods.

Current law in both Europe and America is a muddle when it comes to gene– and cell-based patents. In the USA patents are much more easily obtained. Once a gene is patented the owner is probably covered for almost all its uses. A company can say: 'You can't work on that gene, it's mine,' and lawyers will make a lot of money defending the patent. This puts great power into a few hands. Since genetic sequences are discovered rather than invented, perhaps the gene sequence should count as 'information' rather than an invention, and as such should never be part of a patent. Possible novel uses of the gene or its products or novel methods by which the gene was sequenced could be subject to patent laws. Commercial companies do have to obtain some protection for the financial efforts that they put into their years of work. They need to recoup their investment and in addition to make a profit. If they cannot have patents the only other resort is secrecy and that is perhaps even worse. Currently firms seem to be putting in patents for gene sequences in vast numbers and the outdated patent laws about what exactly can be covered are in urgent need of clarification.

7

The Family and the New Eugenics

To leave his wife, to leave his babes
His mansion and his titles, in a place
From whence himself does fly. He loves us not;
He wants the natural touch: for the poor wren
That most diminutive of birds, will fight
Her young ones in her nest, against the owl.

—William Shakespeare, *Macbeth*

Some of the great laws of society are derived from fundamental natural principles, which in turn are derived from events associated with reproduction, childcare and death. Those laws relating to parenthood and the family are at its foundation. If you try to defy natural physical laws not arising from our arbitrary conventions or institutions, you can end up seriously damaged. If you try to defy the laws of gravity or thermodynamics you can split your head open after a fall, or freeze to death on a cold night. If you defy the Mendelian laws of genetics you can end up with a damaged child and mother or, even worse, a dead baby.

Government policy in the UK has recently been centring on family law, tradition and a sense of solid family values, with the major aims of providing a stable, secure environment in which to raise children. The aims are to give the next generation a complete and generous education which will enable them to live their lives justly, skilfully and magnanimously. Knowing some of the genes that affect common diseases such as cancer, diabetes, heart attacks and strokes, as well as personality 'traits' such as obesity, neuroticism and alcoholism, may broaden the scope of rules and laws affecting the lives of many more families. Health services, insurance companies, employment offices, industry and family lawyers are going to play a much larger role in decisions that should probably rest primarily with individuals and their families. This chapter will consider some of these problems related to family matters. Because of their complexity, the really interesting questions (how far to use the techniques, who will have access to them and at what expense) will largely go unanswered.

The family: gender selection

Professor Tortula, an eleventh-century Italian physician at the Faculty of Medicine of the University of Salerno, advised that 'if a couple wishes to have a male child let the man take the womb and vulva of a hare and have it dried and pulverised; blend it with wine and let him drink it. Let the woman do the same with the testicle of the hare and let her be with her husband at the end of her menstrual period and she will conceive a male.' Others have it that if juice of cassia, jambolana and veronica is thickened with the excrement of a male monkey and smeared on the woman's yoni she will certainly conceive a male child. However, if you use a paste made of hog-plum, ghee and the sloughed skin of a cobra in my own opinion this will prove completely useless.

Nowadays things are a little more scientific but perhaps just as complex. Parents can now choose the sex of their children without requiring a termination of pregnancy. After *in vitro* fertilisation the embryos can be screened before implantation into the mother to see if they carry male or female sex chromosomes, and the unwanted embryos can be discarded. Alternatively sperms from the husband can be screened before use to see whether they carry a single male or female chromosome, and if a male child is desired only the sperm that determine this (the Y-carrying sperm) can be used. The sperms can be separated by a technique called flow cytometry after labelling with a fluorescent dye to light up the chromosomes inside the sperm.

There are several conditions under which parents may want to choose the sex of their child. Apart from religious or economic reasons, there are legal issues relating to the inheritance of property, descent of title and other issues such as personal preferences. In the USA, gender selection for social reasons is legal but not in the UK. In both countries it can be used to prevent transmission of serious or life-threatening disease that is carried on the sex or other chromosomes. These diseases include cystic fibrosis (where the lungs and other organs fill with sticky secretions), some forms of premature blindness (including retinitis pigmentosa), and Tay-Sachs disease. Gender selection for other than serious disease is legally banned in the UK because it is said that majority opinion is against it. If parents were to be freely allowed to choose the gender of their child this could lead to a slippery slope, the descent into choosing other physical and mental characteristics for a 'designer' baby. However, it seems unreasonable to put a complete ban on the whole procedure for fear of possibly dire consequences to the future human gene pool that in fact may never come to pass. The ban on gender selection in the UK will no doubt be challenged in the near future on the basis of the European Convention on Human Rights Act. This maintains the freedom to found a family of one's own choice. Indeed, recently, one family, the Masterstons from Monifieth in Scotland, sought permission to choose the sex of their fifth child. They already had four sons and wished to have a girl following the accidental death of their only daughter. They were not allowed by UK law to do this but had to go to Italy to find a doctor willing to help them. After fertility treatment, the only viable embryo turned out to be male. They decided not to go ahead with implantation.

In China an extreme form of gender selection has developed recently as a result of a policy of restricting each family to having only one child in order to control population numbers. Transgressing this law can lead to extreme penalties. One healthy baby was recently drowned in front of its mother by officials in the village of Caidan, in Hubei province, because of a violation of this rule.

Strong social attitudes to gender go back a long way in China. In AD 250 Fu Hsuan wrote: 'How sad it is to be a woman. Nothing on earth is held so cheap. A boy that is born comes to earth like a God. No one is glad when a girl is born.' The poet Po Chiu-I (AD 772–846) in the poem 'Remembering Golden Bells' wrote on the birth of his daughter: 'Not a boy – but still better than nothing.' For religious, social and economic reasons a boy is valued more highly than a girl. A son is seen as a 'deposit in the bank', whereas daughters, who traditionally leave their family home at the time of marriage with a sizeable dowry to go to live with their in-laws, have for centuries been regarded as an 'investment without a financial return'. Chinese parents practised female infanticide in the past and probably still do so today. They also use selective abortion after sexing their child in the uterus with high-quality ultrasound machines. Doctors who use these machines are forbidden to tell the mothers the sex of their child but easily find ways of doing so. 'I heard of a doctor who would nod if the result showed a boy, but rap the table for a girl,' said a Beijing gynaecologist. Evidence that sex-selected abortion is occurring comes from the rise in the ratio of male to female live births that have risen from 1.08 in 1989 to 1.097 in 1991. Such an imbalance can only be explained by the abortion of female embryos. Alternatively, girls are abandoned or left to orphanages. By 2020 there is predicted to be a surplus of males in China that will exceed the entire female population of Taiwan and most of them will be condemned to perpetual bachelorhood.

A similar situation occurs in India where boys are considered socially and economically at a premium. Although an act passed by the Indian government in 1994 outlawed the selective abortion of female embryos the practice is still widespread. At a cost of often less than 500 rupees (about £7) parents can determine the sex of their unborn child and then go on to termination if it is female. In one study reported by UNICEF in 1984, 8,000 foetuses were aborted in Bombay (Mumbai), of which 7,997 were female. In another study of a Pakistani hospital in 1985, 700 women sought prenatal determination of sex and all the male embryos were kept for birth, whereas 430 out of the 450 female embryos were aborted.

The rise in the ratio of male to female live births is occurring all over India but especially in the northern states such as Bihar and Rajasthan where the ratio has reached 1.07, one of the highest in the world.

Passing on bad genes

The genes on the female X-chromosome are not particularly bad, but could a child of the future sue its parents for handing on a set of disabling genes? Before the choice of embryos was possible such a lawsuit would make no sense, since the parents had no control over which genes were being transmitted. But now it could come into the same legal category as a tort, whereby actions are taken against the mother for harming her unborn child. Such actions have already occurred against mothers who take recreational drugs during pregnancy, since these may harm the developing child. In 1998 Cornelia Whitner was sentenced to eight years in prison in South Carolina. Her crime: she ingested crack cocaine whilst pregnant, and could have harmed her unborn child. Fortunately she later gave birth to a healthy child. The State Attorney-General, Charles Condon, proclaimed: 'A viable fetus is a citizen and fellow South Carolinian'; the mother was sentenced by the Court under the State Child-endangerment Statutes of 1985. The law in South Carolina contradicts the Supreme Courts in five other states which have dismissed criminal charges against pregnant women whose behaviour could have harmed their unborn child. This included a bizarre case of a pregnant teenager who shot herself in the abdomen; her baby died but she survived. She was acquitted of murder. But it is possible that in future parents will be held responsible for failing to repair genetic defects at the embryo stage of their children. 'Wrongful life' or 'wrongful birth' lawsuits could ensue.

'Wrongful life' lawsuits have already been brought by parents against doctors, but not as yet by children against parents. One such case involved Paul and Shirley Berman whose daughter was born with Down's syndrome. The parents claimed negligence on the part of the doctors in not advising them to have a chromosome test, despite the fact that Mrs Berman was thirty-eight at the time of the pregnancy. The parents brought a 'wrongful life' suit against the doctors, seeking compensation for the suffering their child would experience during her life, as well as bringing a claim to compensate them for their own emotional anguish. The New Jersey Supreme Court rejected the Bermans' claim of 'wrongful life', stating that they could not judge 'the difference in value between a life in an impaired condition and the utter void of non-existence'. Justice Morris Pushman wrote: 'Ultimately the infant's complaint is that she would be better off not to have been born. Man who knows nothing of death or nothingness cannot possibly know whether that is so.' However, the Supreme Court did award damages to the Bermans for their emotional anguish as part of their 'wrongful birth' claim. With the newer techniques of genetic screening for embryonic disease the issue of parental responsibility will come increasingly to the fore. It is only perhaps a matter of time before parents will be held responsible for failing to correct the genetic defects that they pass on to their children.

Who is the father (or mother)?

The new eugenic techniques are leading to much confusion and contradictory legislation in attempts to establish an answer to the simple question of who are, or should be, the parents of one particular child. Two fathers dispute the custody of a child: one is a wealthy but unsavoury character; the other a poor but sensible man who already has a large family. An old-established way of settling the problem and a governing principle under English Law is to decide in the best interests of the child. Does this mean a comfortable, materialistic lifestyle under the influence of an inveterately stupid father? Or poverty, poor educational opportunities but with a reasonable father? This requires the judgement of Solomon, who, after all, had experience in these matters; even he might find it quite difficult to decide. There can be no hard and fast rules about what constitutes the child's best interests. If one wanted a more scientific judgement of paternity it could be resolved by establishing who is the genetic father and awarding him the custody.

Paternity testing to decide the genetic father of a child is now common practice. A 'fingerprint' of DNA reveals all the genetic variations, that is all the different mutations that are found in the father's genes, and can be used to see if it accurately matches the variations in the child's DNA. If it does, paternity is proved, it being extremely unlikely that the match has come about by chance. Advertising posters have now sprung up in more than thirty cities in the USA asking passers-by, 'Who's the father?' A genetics company, Identigene, will obligingly give you the answer after taking cells from the mouths of mother, child and the alleged father. This will, say the company, give the family 'peace of mind'; this may be rather short-lived, however, because from 10 to 20 per cent of children in previous studies have been found to have different genetic fathers from those whom they believed to be their own. However, the test can be used in happier circumstances to confirm that the newborn child that you proudly take home from hospital is really your own and not someone else's. The tests have also been used to settle legal disputes about inheritance of assets to a biological relative, such as father to son, when the legitimacy of the son has been contested.

In the past any child could have reasonably expected his father to be present at the time of conception even if he did not turn up for the birth. But with the new reproductive techniques of surrogacy and sperm donors the situation can become quite complex. A child can now have four different 'fathers'. If a couple decide to commission another woman to carry their child as a surrogate mother the resulting child can have 'fathers' from (1) the husband of the commissioning couple; (2) the husband of the surrogate mother; (3) the husband of the egg donor; and (4) the man who donated the sperm if different from (1). Who then is financially responsible in law for the support of the child? The Family

Support Agency has been somewhat confused. There is a growing tendency to allocate social and legal responsibility to the genetic father; but this would not work if the sperm were derived from a donor who remains anonymous.

However, considerable social and financial responsibility is allocated to the social father by divorce laws which confer duties on him even if he is not the genetic father. These laws seem to be trying to cover all the possibilities without any clear policy. Perhaps there are no 'eternal principles of justice and morality' in this field. Both the European Convention on Human Rights and the UN Convention on the Rights of the Child contain statements about the individual's identity rights. Article 7 of the Convention on the Rights of the Child specifies that a child has the right 'as far as possible, to know and be cared for by his parents'. The Data Protection Act of 1984 demands respect for the private life of an individual. In the case of a pregnancy resulting from an anonymous sperm donor does this mean that the parents have a right to keep this a secret, or does it mean the child has a right to know who is his genetic father donating the sperm? The field is full of such dilemmas.

In 1997 a court decision in California added even more confusion when it judged a child to have no legal parents at all. The case arose when an infertile couple, Luanne and John Buzzanca, contracted three separate adults – a sperm donor, an egg donor and a surrogate mother – to bear a child for them. A baby girl called Jaycee was eventually born. Before the birth took place John, however, left his wife and filed for a divorce. Luanne took custody of the baby after the birth and sued John for child support. John argued that he did not owe the child anything since the baby was not a 'child of the marriage'. The court was naturally bemused by the problem. The baby girl had three biological parents and a social mother, yet the court managed to rule that the child had no legal parents. Luanne was not the mother because she neither contributed the egg nor gave birth. John could not be the father because he had no biological or social relationships with the child. The surrogate mother had forgone parental rights in the surrogacy agreement. This meant that no one was legally responsible for the care of the child. If the child were unlucky enough to be born with severe medical problems such as a cerebral palsy no one would be obliged to deal with this.

Naturally there was an appeal against the decision. In March 1998 a three-judge panel ruled unanimously that both John and Luanne were to be considered the legal parents because the child 'never would have been born had not Luanne and John both agreed to have a fertilised egg implanted into a surrogate mother'. John had caused the child's 'conception every bit as much as if things had been done the old-fashioned way'. The court therefore went beyond the traditional views of genetic heritage or social milieu establishing paternity, adopting the view that 'intention to have a child' was sufficient to qualify a man for financial support of 'his child'.

There is more confusion and ambiguity regarding who is the mother. Now that it is possible to store frozen eggs for long periods, some mothers are depositing eggs in storage for future use by their daughters unfortunate enough to have been born without functioning ovaries. So if their daughters want to become pregnant with a genetically related child in the future they can use these eggs for *in vitro* fertilisation with their husband's sperm. This would mean any subsequent child they bear would be in fact be their half-sister or half-brother rather than a daughter or son.

What actually constitutes the relationship of motherhood? Is it to be defined in purely genetic terms as the source of the maternal X-chromosome? Or are physical, emotional and social ties of more importance? This became an issue in a custody dispute in California, *Johnson v Calvert* (1993). The Calverts provided both eggs and sperm but contracted Anna Johnson to become the surrogate mother and she carried the baby to full-term delivery. Anna Johnson then refused to give up the child since she considered herself to have the true feelings of the mother. The judge awarded the Calverts sole custody of the baby but the reasons he gave were not on the basis of the original surrogacy contract; nor were they, as traditionally, in the best interests of the child. This would have worked because Anna Johnson was a single mother, whereas the Calverts could clearly provide a better home for the child. Instead the judge justified his decision on the basis of the child's genes. He noted the 'tremendous need out there for genetic children', and called Anna Johnson, the surrogate mother, a 'genetic hereditary stranger' to the child. He considered the source of the genes to be the primary factor determining the maternal relationship and not the physical and emotional bonds that might have formed. The conflict of 'genetic rights' versus emotional and social bonding comes into focus in many such custody cases and the law can be very ambiguous about the issue.

In another case a white couple from New York unexpectedly gave birth to a black baby when the mother was delivered of male twins following infertility treatment. There was no past history of Negroid ancestry in either parent, so the only reasonable explanation was that she had been mistakenly implanted with the embryo of an African-American couple whilst she was undergoing infertility treatment. This mix-up of embryos has led to the white couple filing a malpractice suit against Dr Nash, their fertility doctor, but the problem still remains of what to do with the black baby. The white couple do not wish to raise the child and would like to give custody to the genetic parents if they could be found. Lawyers working on the case hope someone will take responsibility for the black child even if the genetic parents do not want to themselves. They also hope that a visitation agreement can be made so the boys will grow up knowing each other.

In Britain, the case of *Blood v Human Fertilisation and Embryology Authority* illustrates another of the pitfalls into which ambiguous family legislation may fall when all future circumstances cannot be anticipated. This case involves a conflict between the common right of any individual to found a family and the dislike by lawyers of the use of semen for fertilisation without the donor's signed consent. The case arose when Mrs Diane Blood, a widow, requested that the UK Human Fertilisation and Embryology Authority be allowed to use her husband's semen to start a pregnancy. The couple had been married for four years and had been trying to have a child for two months. The husband died suddenly from bacterial meningitis and while he was on the life-support machine doctors removed a sample of semen for storage. This was obviously done without his consent but at the request of his wife. Mrs Blood was later refused use of the sample because there were no signed consent forms. Her late husband was categorised by the Act as a non-consenting donor. When she applied to go abroad with the sample the Authority, which had the discretionary power to permit export of the dead husband's sperm, again refused on the grounds there was no donor consent for the export. For them it would be wrong to enable a procedure deemed illegal in the UK to be done elsewhere. This could be seen as undermining the will and dignity of the Parliament of Great Britain. If the dignity of Parliament is so tied to whether Mrs Blood has a baby, such dignity seems to be a terrible encumbrance. In the view of the Human Fertilisation and Embryology Authority, it seems, the dignity of Parliament is more important than allowing a woman the common right of being able to bear her husband's child. Mrs Blood lost her application after a judicial review of the decision by the High Court, whose power was limited to determining whether the Human Fertilisation and Embryology Authority had acted within the scope of the rules, and found, inevitably, that it had done so. Mrs Blood was left in the anomalous position of being able to have a sperm donation from an anonymous man but not from her late husband.

Persevering in this absurdity to preserve dignity, a later ruling by the Court of Appeal allowed Mrs Blood to use her late husband's sperm for insemination provided she went to a European Community state where the procedure was not illegal, such as Belgium. She eventually conceived and gave birth to a boy weighing 5lb 3oz – a happy outcome. Further legal pettiness arose when she was refused the right to put her late husband's name on the child's birth certificate. Babies conceived posthumously are said to be legally fatherless, as set out in the 1990 Human Fertilisation and Embryology Act. This appears to make nonsense of one of the fundamental laws of heredity that require both a paternal and

maternal set of chromosomes for any child. This ruling may well
be due for reform in the near future.

The general issue raised of taking and using semen or eggs
without consent of the donor and the wisdom of posthumous conception are not to be dismissed lightly particularly as the resulting
child would be considered legally to be fatherless. Under common
law, no medical procedure can be carried out on an incompetent or
comatose person, like Mr Blood, unless it can be justified as necessary or in the patient's best interest. So the legality of the doctors
taking semen in the first place was questioned. But the laws relating to this and other eugenic procedures are meant to be a
repository of collective reason, common sense and equity. If the
rules and regulations are tied so tightly, many people will take up
the fight against them. The couple had been trying to conceive for
two months before the fatal infection occurred. Lawyers claimed
that Mrs Blood's subsequent desire for a child may have stemmed
from persistent grief, or that she just wanted a 'souvenir' baby.
This would not be in the best interests of the future child. How
could they know that? It would not prevent her from using an
anonymous donor instead of her late husband. Allowing Mrs
Blood to do the latter would in no way harm or endanger any
other member of society or set an undesirable precedent, except
possibly depriving a baby of a legal father (a law that could be
easily changed). Other countries think differently. In the USA in
1991 a number of wise soldiers had banked their precious semen
for posthumous conception before the Gulf War. If they were to
be killed at least they could procreate by proxy and their name
would last.

Laws do not make people any more ethical. But if the ethics
remain intact they can correct the vices of bad laws. A bioethics
authority for the future should not make legislation that is so rigid
that it produces lengthy, distressing and expensive court procedures; it is estimated that Mrs Blood's case cost her more than
£100,000. It should not appear to deal unjustly with particular
cases such as this one. Mrs Blood has now had her child,
Parliament is still standing, and no one's real authority has been

undermined by this exercise. But at what cost for the loss of generosity, humanity and dignity?

The case of Diane Blood illustrates two further points: first, that doctors are trained to consider the problems and needs of an individual under a particular set of conditions. Lawyers try to follow to the letter the legislation for a category of person, not taking into account particular conditions. The law tries to treat everyone equally, regardless of differing circumstances. Second, the case raises the possibility of procreative tourism, particularly within the European Union, where free movement to obtain services unavailable in the country of domicile is central to the notion of the Union. Another example of such tourism is embryo splitting, which is illegal in the UK, so couples have to go to Italy if they want to store split embryos for future use. This suggests that domestic regulation should, so far as possible, be based on standards and principles agreed within the European Union. The outcome of a series of individual court cases in different countries covering a range of controversial issues may be the best way of determining what is finally acceptable to the Union.

Selecting embryos

Another difficult problem for the future will be a conflict between parental requirements and established legislation as to the extent to which designer babies will be regulated. The mutations listed in Appendix Tables 1 and 3 can all be screened in pre-implantation embryos and if any are found the embryo could be discarded and another one used in its place. This is currently being used to prevent the transmission of serious inherited disease such as cystic fibrosis. Although pre-implantation diagnosis avoids the need for selective abortion a major disadvantage is that it requires *in vitro* fertilisation. Currently this only has a success rate of about one in five although rates are predicted to improve markedly in the next decade. There are other problems with the procedure, as described on page 40. The use of such techniques by married couples to select for the presence of particular characteristics in healthy embryos will need to be critically evaluated. It is natural that parents would want, and have the right, to choose the best possible health for their children. Pre-implantation screening would be one way of selecting which of their genes to transmit to their offspring.

One recommendation that the European Commission made for the regulation of prenatal diagnosis (of which pre-implantation diagnosis may be considered a special case) is that prenatal diagnoses should always be considered as a medical act and be offered on the basis of specific medical indications. The choice of sex or other characteristics for non-medical reasons is an ethically unacceptable indication for prenatal diagnosis and should be prohibited. But even the straightest of ethical issues can start to look oblique on closer scrutiny. Most of them end in ambiguity. The term 'medical indication' is an example of this. We are led to think of disease as an isolated disturbance in a healthy body, but some diseases could just be phases of certain periods of bodily development, for example the bodily deterioration associated with ageing. Are alcoholism, obesity or anxiety states medical diseases or undesirable character traits? Would they come within the scope of prenatal screening if genetic markers were available for diagnosis of these conditions?

New developments

There is now a raft of new and ever-expanding techniques that combine the new information of genetics with the practical advances in assisted reproduction, to obtain the 'best' outcome for the newborn child. The issue now is who will be able to use these new techniques? Will it be freely up to individual choice, or will the state and other agencies exert more and more control over the use of the technology?

The older methods of sterilisation and abortion are now still mainly subject to individual choices, as are the newer methods involving sperm banks, *in vitro* fertilisation and surrogate motherhood. The latest techniques for storage of eggs, embryo selection and designer babies are now subject to much greater outside control. It appears that we cannot believe in or rely on the intelligence and decency of the majority of men and women. They are not free to do as they please with these techniques. And it is possible that the future methods such as embryo enhancement and cloning will come under even stricter controls. However, freedom to use the older methods is actually more apparent than it is real. Increasing pressure is being exerted on individual couples by industry (persuading them to purchase its eugenic products); by insurance companies (that want to match premiums to health risks); by employment agencies (that want to minimise the health hazards to their workforce); and by government commissions (for a variety of reasons such as the desire to reduce crime rates, alcohol or drug addiction and wanton aggression or other more biologically determined offences). The obvious dangers are that the newly developed eugenic techniques that are initially meant for free and voluntary use by individuals will be increasingly taken over by outside agencies.

Just one example of the sort of thing that can happen: a pharmaceutical company, Roussel, has developed a contraceptive called Norplant that, after surgical implantation, will prevent pregnancy for five years. It was approved for public use by the US Food and Drug Administration in 1990. As a form of long-acting contraception Norplant appealed to many different groups of women: it is a very effective, passive form of birth control that needs no attention for five years. Its very properties, however, make it tempting for politicians to use as a means of controlling reproductive rates in certain classes of 'undesirable' women. A poll in the *Los Angeles Times* in 1991 recorded that 61 per cent of respondents approved of enforced use of Norplant among those women of childbearing age who were also abusing drugs. Between 1991 and 1992 thirteen state legislatures introduced bills that proposed offering Norplant to women on welfare benefits as a way of avoiding unwanted pregnancies, and further offered them financial inducements to take this form of contraception. All these bills failed to get through, albeit by narrow margins. Kansas, for example, proposed that women choosing the programme would receive benefits of $500 at the start and thereafter $50 annually, provided the implant remained intact. The aim of the programme was not strictly eugenic but had obvious eugenic overtones: it was more 'to save the taxpayers millions of their hard-earned dollars'. A judge in Visalia, California, ordered a mother on welfare benefits who also abused her children to use Norplant as an alternative to a longer jail sentence. He argued that the 'compelling state interest in the protection of the children of the state supersedes this particular individual's rights to procreate'. The *Philadelphia Inquirer*, among others, supported such judgements and published an editorial entitled 'Poverty and Norplant – can contraception reduce the underclass?' The reasoning was that more African-American children live in poverty because African-American women, having the most children, are the ones least capable of supporting them. African-Americans immediately attacked the editorial as racist and as a thinly disguised form of eugenics. It appeared to some to be a resurrection of the old-fashioned eugenic ideas aimed at improving society by temporary sterilisation of a class of 'undesirable' women on low incomes. However, one can foresee how governments of some countries might take an interest in this form of contraception because of the problems of over-population that threaten the country's resources. India has unsuccessfully tried to reduce its population explosion by giving men easy access to vasectomy. China has been more successful with its policy of one child per family. Population control is likely to become a major ecological issue for many countries in future, particularly when the problems of ageing have been overcome (see page 239) and an individual's life span could increase to well beyond a hundred years. Government control of reproductive rates may then become both necessary and legitimate.

8

Society and the New Eugenics

> If a man takes no interest in public affairs, we alone do
> not commend him as quiet, but condemn him as useless.
>
> —Thucydides, *The Peloponnesian Wars*

Immanuel Kant (1724–1804) ended his book *The Critique of Practical Reason* with these words: 'Two separate things fill the mind with wonder: the diversity and design of the starry heavens above me, and the internal moral laws working within me.' The history of eugenics has deeply offended the latter. The subject appears to cry out for the need to regulate the new and powerful social applications that are being developed. Society will expect rules and laws to protect it from the excesses of overenthusiastic doctors and scientists, from the greed and ambition of corporations and profiteers, and from political manipulators out to gain or keep power for themselves. Research doctors are often involved at the developmental stages. But we are not easily prone to professions of humility and too easily develop bees in our bonnets about special theories of disease and treatments. There is admittedly an enormous difference between a good doctor and a bad one, yet sometimes there is little difference in outcome for the patient treated by a good doctor or by no doctor at all. There are some patients whom doctors cannot help, but there are none whom they cannot harm.

Society's response to the new developments is not going to be easy. There is a variable and fine line between law and ethics; that line separates what can be left to individual choice and what requires enforcement by the law. The line is never fixed: '*autres temps, autres mœurs*'. The rules of professional bodies at present relating to eugenics are full of inconsistencies. One problem is that major public interests such as education, health and welfare, quite properly the concerns of law, are so intertwined with the private and individual interests of having a family. The latter could be left to individuals to work out for themselves on the basis of their own wishes, needs and preferences. For the future we will have to balance the practical applications arising from the new research against the need to protect both individuals and society from its dangers, then to resolve in some way the many conflicts of values that will arise; for example, how far to allow the development of designer babies.

Law is needed to reconcile fact and value and should be firmly based on reason and common judgement. Unfortunately, the facts and techniques are changing from month to month. In the UK the Human Fertilisation and Embryology Authority developed a set of rules to cover the use of human embryos produced by *in vitro* fertilisation, but these rules do not specifically apply to embryos produced by nuclear transfer techniques. Up to 1999 there was no UK legislation in force specifically directed at human cloning. When human cloning was banned in the USA, laboratories there started to use empty (enucleated) egg cells from a cow for transfer of a human nucleus to make cow–human hybrids. So the law will have to be amended again; just as quickly the scientists will find a way round it.

The problem is likely to get worse with the constantly increasing demands for new treatments of infertility. This is partly because there is a continuing fall in the number of babies available for adoption and a growing tendency for women to delay having children, electing instead to concentrate on their careers. Once they decide to have children they are often at an age when problems of conception and childbirth are more likely to arise. One in six couples now has fertility problems in the UK. In Western societies the number of women undergoing *in vitro* fertilisation rose from approximately 12,000 in 1992 to 19,000 in 1994.

Legislators do not always understand the intricacies of the science they are called upon to regulate, so regulatory agencies composed of a small group of scientists, lay people, and lawyers are set up to provide recommendations and guidelines. These regulatory bodies have proliferated to such an extent that they now form a mosaic of ill-assorted agencies. They range from local hospital ethical committees and national committees such as the UK Human Genetics Commission, the Advisory Committee on Genetic Modification and the UK Human Fertilisation and Embryology Authority to international bodies such as the European Medicines Evaluation Agency. It is not surprising that with the numbers involved they are already starting to give confused or contradictory advice. To give one example: the UK Alzheimer's Disease Genetics Consortium has drawn up guidelines for what it considers to be the best possible practice for the genetic diagnosis of Alzheimer's dementia. It proposes that before genetic testing is undertaken informed consent should be obtained from the patient, but the results should not usually be given back to the patient, except under the most exceptional circumstances, and then only after appropriate counselling. However, another body, the Association of British Insurers, has demanded as part of its genetic policy that for insurance policies of more than £100,000 the results of genetic tests should be declared to it. Their agenda is to try to prevent adverse selection (as discussed on page 144). A third body, the UK Human Genetics Advisory Commission, recommended a moratorium for two years on the use of genetic information for insurance purposes until its actuarial relevance is evaluated.

Additional problems of leaving such policy decisions involving the genetic technology to various consortia are that wider issues may be overlooked. Genetic screening for a particular protein variant has been in use as an important diagnostic aid for blood fat disorders in hospital clinics worldwide for the past few decades.

This information, available in case notes and medical reports, can now be used to predict the risk of occurrence of Alzheimer's dementia in some of the patients who were tested. There is no clear formulation on the confidentiality of these records or how they should be used. Who should have access to the information? The patients were initially tested for a different condition, namely a cholesterol transport disorder. Should they or their first-degree relatives now be told of the calculated risks of developing a dementia? Should their primary care physician be informed of the results? Or should the results just remain undisclosed in the hospital records? There are no obvious answers.

In this chapter some of the questions and problems to which regulation (or legislation) can give rise will be discussed. The answers will necessarily be ambivalent, and any solutions will not be easy to implement. But the responses can vary according to whether professional ethical codes, moratoria, referenda or legislation are deemed appropriate.

Industry

By the year 2010 commercial developments derived from the Human Genome Mapping Project may account for sales of $60 billion worth of products per annum. That is half the international sales of the pharmaceutical industry in 1992. New genetic knowledge translated into products will make a profit in almost any marketplace. Avarice and the delusion that one must do as well as one can for oneself are the great spur to industry. Those who control such sales stand to make enormous financial gains in the next few decades.

Above all, satisfying the shareholders appears to be a higher priority than allowing for sustainable development and supporting the needs of the local population. After all, the financiers, without much care for the importance of ecology, have successfully exploited the riches of the globe – the forests, the sea and the metals. So why not now exploit the human gene pool? One molecular biologist, Herb Boyer, invested $500 to help launch the biotechnology company Genentech. When the company was floated on the stock market his stake was worth $82 million, giving rise to a newspaper article entitled 'Science – for love or money?' Such is the fatal dance of the dollar, with prices zigzagging irresistibly up on the stock markets. Shareholders demand – and rightly so – the creation of value as reflected by increasing profits and stock prices of the companies in which they invest. A sign of their real wealth is to be able to buy paintings before asking prices.

But the widening gap between the profits of the multinational companies and the economic impoverishment they can bring to local communities, and the damage to the environment of developing countries where they operate, is a serious drawback. The global corporations should realise that when Jehovah parcelled out the Earth it was not meant only for the likes of Coca-Cola, Ford Motors, BP-Amoco and IBM. Apart from creating wealth, big corporations need to ensure that the globalisation of their activities delivers more than a litany of dashed hopes to the developing societies in which they work.

Naked self-interest and the callous profit motive are both starting to threaten and drown the spirit of collaborative enthusiasm and cooperation in the fields of genetic research. To give just two examples: there is a primitive tribe of hunter-gatherers, the Hagahai, living in a remote and inaccessible mountain range of Papua New Guinea. Epidemiological studies suggested that some members of the tribe were making antibodies to a virus that can cause a form of leukaemia. Cells were taken from a young male member of the tribe and cultured in the laboratories of the National Institutes of Health (NIH) in Bethesda, Maryland. The cells were indeed found to be making antibodies and NIH researchers were quick to see the profits to be made out of them. They applied for patent rights relating to the cell cultures they intended to establish. It was agreed that the Hagahai should receive some (unspecified) share of any commercial profits arising from the 'invention'. In 1996 the patent was officially disclaimed by the NIH after heated international controversy developed about the exploitation and biocolonisation of primitive peoples.

In the other case a biotechnology company, Sequana Therapeutics of La Jolla, California, financed a team of scientists to visit Tristan de Cunha, a remote volcanic island in the Atlantic which is home to several hundred descendants of the British who landed there in 1817. Why they are interesting now is that about half of them suffer from asthma. Sequana scientists took blood samples from about 270 residents and later reported that they had found two candidate genes responsible for asthma. However, they refused to share their findings with the rest of the scientific community until they were in a position to exploit their observations commercially, at the very least not before patenting the genes in question. They were accused of putting business considerations before collaborative efforts to find a useful cure for the disease. They made no bones about their motives, saying they were in the business to make a profit out of their work, which they could only do by keeping their results to themselves. After making their profits they could then perhaps divert some of them back into further research as well as pay for their initial expenses in setting up the research project.

It can be seen from both these cases that the cells and genes of remote rural peoples are now becoming the intellectual property of corporations and government institutions in the hope of filling the cash boxes of the research and genetic laboratories. The slick virtuoso tricks of the globalised industries are appropriating the genetic raw material of indigenous peoples to create considerable wealth for their shareholders (and naturally for themselves). They claim in their defence that free market incentives and the profit motive are the best and most efficient ways to advance certain types of basic medical research that can be very costly.

Nearer to home, Iceland has an isolated population with very little immigration for the best part of a thousand years. It is therefore an ideal place to investigate the genetic basis of disease. In December 1998 the Icelandic Parliament passed a bill making it legal for a private company, deCODE Genetics, to have access to a comprehensive health care database containing the very detailed medical records of almost the entire 270,000 population of

Iceland. This database, combined with genealogical records that are already publicly available, is of vital interest because it contains large family trees for the vast majority of the population, the records having been kept meticulously up to date.

deCODE Genetics, largely funded by American investors, has signed a deal worth about $200 million with the Swiss pharmaceutical giant, Hoffman La Roche, giving the company exclusive access to parts of a newly established DNA database. The company will then be able to link it with the other two databases to search for the genetic origins of some twelve common diseases. It could be considered that the entire population of Iceland is being turned into a biomedical research commodity with a large commercial potential. The DNA part of the database is now being collected and will cover about 90 per cent of the Icelandic population.

Opposition to the bill was based on three major arguments: (1) there would be no individual signed consent for transfer of the health information to third parties and this would constitute an unwarranted invasion of privacy (provision for an opt-out system was made, but this was difficult to access); (2) it may lead to loss of trust in the confidentiality of the doctor–patient relationship even if the health information were to be encrypted before the transfer; (3) a monopoly was being given to one company to pursue this research. The project will cost about $150 million – more than twice the entire research budget for Iceland. So proponents argue that a private company is needed to fund the project since no governmental agency would have the marketing experience or the willingness to float such a large venture. The bill was finally passed and the DNA database is being collected.

Such accurate and extensive medical records combined with a matching DNA bank could provide a cornucopia of information to help uncover the genetic origins of many diseases for the inestimable benefit of the Icelandic as well as other populations. But if the database were to fall into the wrong hands it could provide yet another means for discrimination against an 'undesirable' or minority element in society. In passing the bill, the Icelandic Parliament also broke one of the fundamental principles of scientific research as formulated in the Helsinki Declaration of 1964. This spells out the guiding principles for conducting research involving human subjects and includes the proviso that any research projects have to be evaluated and approved by an independent ethics committee before recruitment of patients can begin. deCODE managed to convince the Icelandic government to pass the laws, which bypassed the necessity for review by a bioethics or data protection committee. The government claimed that the act resulted from an informed democratic decision, but only 13 per cent of the nation considered themselves to have a good enough grasp of the nature of the bill, according to a Gallup poll conducted in November 1998.

This raises an interesting issue that obtaining informed consent for genetic studies assumes that the person has a full understanding of the subject matter. This further implies that the person should have a duty to know all about his genetic constitution and the methods to be used before proper informed consent can be given. A very unlikely state of affairs.

The Act was passed through the Icelandic Parliament before the public really understood all the issues. A provision in the bill ensures that the entries to the database will remain anonymous and therefore should not need informed consent of the participants. This then raises the problem that if some Icelanders are found to possess genetic variants that predispose them to a serious but treatable disorder they will not be told about it. Yet it is the doctors' duty to inform patients in such circustances. Clearly the database is not going to be as anonymous as was initially promised. The more genetic technology that is introduced into society, especially by industry, the more important it becomes to protect the rights of individuals from either becoming subservient to state procedures or becoming exploited by market-led commercial forces.

Home tests

Do-it-yourself tests for detecting pregnancies or testing for paternity have been a recent commercial success. The discovery that genetic variants can predispose to some types of inherited forms of cancer has provoked widespread public interest. Demands for home tests by the public have been partly satisfied by commercial laboratories. Tests for genetic variants that cause a rare cancer of the large bowel have been put on the market and, after referral by a doctor, are being performed by a private American company, La Corp of Baltimore, Maryland. Many inadequacies and errors, however, were found in the tests carried out. In a study conducted in 1995, 177 patients were tested but only 19 per cent received genetic counselling before the test to explain what was going to happen. Only 17 per cent provided informed consent to have the test. In 32 per cent of the cases the referring doctor misinterpreted the test results. Some patients at risk of developing the cancer were given false negative results because the tests were not done on affected family members to see which particular genetic variant occured in that family. The scope for misuse of the tests – misinforming the patients and the creation of public anxiety in view of the other variables that can influence the final test results – are enormous.

Other commercial gene-based tests are also appearing. One developed by Genentech in California has received recent approval by the US Food and Drug Administration to be used for predicting the malignant state of a breast cancer. About one-third of patients with breast cancer have a genetic alteration that causes increased synthesis of a particular cellular protein (called Her-2). Such people have a more virulent form of the disease and will respond to different therapies (such as treatment with Herceptin) than will the other two-thirds of patients who have a less aggressive form of the cancer.

Another development, by Myriad Genetics of Salt Lake City, Utah, markets tests for genetic mutations that predispose women to breast cancer. This is useful knowledge to have since such cancers can be treated successfully if caught early. But regulation and tests of quality control for the use of these measurements in the marketplace are going to be required. The need is to protect the public from errors and the generation of unnecessary anxiety.

Life and health insurance: disease prediction

Some people – gamblers, sexual adventurers perhaps – enjoy taking risks. Not so many people enjoy taking direct risks with their health or life – bungee jumpers, whitewater rafters and free-fallers apart.

In its simplest form life or health insurance is a way of relieving the effects of chance events such as accidents or disease. Pooling the modest premiums from a large number of people who are at the same risk can provide resources for making substantial payments to those who suffer harm or damage. The escalating costs of medical care are now making it imperative that private hospitals have recovery rooms adjoining their cashier's office, as well as rooms close to the surgical operating theatres. The blunt truth is that most people just cannot afford to pay unless they have some form of health insurance.

The issues discussed in this section are socially very important. The bottom line is that the extensive use of the new eugenic techniques may raise the need for a national insurance service, similar perhaps to the National Health Service, to provide insurance cover for a developing genetic underclass.

Standard insurance forms represent voluntary contracts between an individual and the respective insurance company. The company asks for details about recent medical treatment, medical conditions and family medical history. They may require medical examination or consent for access to medical records. Assessment of risk plays a part in making the insurance contract. It is for the proposer to disclose information of the possible risks that he is exposed to, otherwise to withdraw from the contract. Failure to disclose relevant information is likely to lead to rejection and a void contract in case of a claim.

There are several ways that health or life assurance could be organised. At least three types of models could be envisaged, based on solidarity, mutuality or altruism. Insurance based on solidarity takes no account of the different levels of risk that individuals bring to the pool. Premiums are set at a uniform level, or based on the ability to pay; this model does not offer much of a business case. The British National Health Service is a good example of this type.

Insurance based on mutuality relates the premiums to be paid to the level of risk each person brings to the common pool. The premiums will therefore vary with risks involved. A car driver with a clean driving licence pays a lower premium for car insurance than a driver with a poor record. This is perhaps fair because the bad driver could have taken more trouble to avoid the accidents. Conversely, penalising someone with higher premiums through no fault of their own because they have inherited some faulty genes seems unfair, just as it would be for financially punishing someone for the colour of their skin. From birth onwards everyone has the potential for a variety of afflictions to match the defects in their DNA, but not all of these will come to light. So calculation of such risks can be difficult, and risk prediction for disease in life and health insurance may draw heavily on the other health details of the person.

Finally, insurance based on altruism argues that an individual should not be penalised for the chance inheritance of unfavourable genes over which he or she has no control, and that if anything health insurance as an instrument of social justice should be made cheaper, not more expensive, for them. This would be a non-commercial operation and would probably need the government to provide some sort of support system.

In the UK most life assurance is commercially organised and based on mutuality, whereas most health insurance is organised by the National Health Service and based on solidarity. Life assurance means that the insurers try to adjust the premiums to match the risks involved. For single gene disorders such as a muscle wasting disease (Duchenne's muscular dystrophy) or cystic fibrosis, this is not difficult to do since, if the individual possesses the bad genes, he will almost certainly go on to develop the disease. The usual underwriting practice of insurance companies is to accept people with a family history of, say, Huntington's disease who have not been genetically tested but to charge them higher premiums. People known to carry the gene for Huntington's disease are usually refused cover (although they can obtain considerably better annuity rates if they retire because their life expectancy is so reduced).

However, for multiple gene disorders such as heart attacks, diabetes or high blood pressure, the calculated risks are far more uncertain. Clusters of genes are involved, not in causing the disease but in conferring a predisposition or susceptibility to the disease. Prediction of risks will be uncertain because of two major factors: (1) the disease will often not become apparent unless certain environmental conditions are encountered; and (2) there is a variable gene-to-gene interaction that can modify the severity of the disease. For example, a woman carrying the susceptibility genes for diabetes mellitus may not manifest the disease until she becomes pregnant. The pregnancy in some way tips her into the diabetic state, her blood sugar levels rise and she starts to pass sugar in her urine. After the pregnancy the diabetes appears to go away, but it may recur at a later date as the woman ages or becomes obese. As an obese person with diabetes she may sometimes see the disease disappear if strict dieting results in weight loss. How then should her premiums be adjusted if the declared aims are to take risks into account? As a general rule the risk is calculated at the time of taking out the insurance. If the individual is not apparently diabetic at the time then no increased premium is applied.

The insurance industry in the UK has already laid out guidelines for the disclosure of genetic tests before the purchase of life insurance, disability income insurance or critical illness cover. The code of practice of the Association of British Insurers (the professional body representing the insurance industry) recommends that for purchase of life insurance up to a total of £100,000 which is directly linked to a new mortgage, the results of any genetic tests must be reported but may not be taken into account by the insurance company. It is implied that for sums greater than £100,000 such genetic tests will be taken into account and the premiums adjusted accordingly. Companies insist that premiums paid even for small policies must truly reflect the extent of risks of disability. Otherwise people at high risk could insure themselves for large sums and stand to make a financial gain out of the premiums paid by the lower-risk individuals. This would be robbing the common pool of people of their money collected as premiums to pay a few individuals who did not honestly declare their risks. This situation is called adverse selection, or 'anti-selection'. Eventually the lower-risk individuals, feeling cheated, may not purchase insurance at all and so reduce the common pool of funds available to supply claimants.

Insurance companies in the UK cannot make applicants take genetic tests but if a test has already been taken the result should be given to the insurance company if it asks for that information. This is one reason why people do not obtain predictive genetic tests for hereditary cancer or other disease, for fear of genetic discrimination weighting the premiums against them when they buy the policy. If, however, a genetic test is taken after the policy has been bought the insurer cannot ask for the results. A woman of thirty with a known mutation in the breast cancer gene (BRCA1) may have a seven-year reduction in life expectation. If she has already had the genetic test the insurers would expect her to pay a higher premium. If the test is taken after the purchase of the policy the premiums are not altered.

However, the new genetic tests for chronic illness such as diabetes, dementias or Parkinson's disease may profoundly affect the purchase of long-term care insurance or critical illness cover by individuals and leave many people unprotected when they come to need most help.

Insurance companies often refuse to insure against pre-existing conditions. Does the possession of a liability gene for, say, Alzheimer's dementia constitute a pre-existing condition? Do such people, although healthy now, come into the category of having a disease? If having a genetic defect counts as a disease, then the whole population would come into a disease category. All human beings suffer from the grave and sometimes fatal genetic defect, the inability to make ascorbic acid (otherwise called vitamin C). Unless we eat food containing ascorbic acid we will

die. But we do not normally think of this inability to make it our-
selves as a disease because we all suffer from it. We all possess
other deleterious genetic variants, but they may never come to
light as disease unless we encounter a particular set of environmental conditions such as lack of a particular foodstuff. It is
difficult to see how insurance companies will deal with this issue.

Risk discrimination by insurers could lead to a social underclass, based on their genetics, who will find it impossible to afford
any form of health or life insurance. Possible long-term consequences of creating such a genetic underclass are described on
page 174 (*Homo technicalis*). It is already happening to some
extent. Epileptics find it difficult to obtain adequate travel insurance and may suffer serious consequences of this if they develop
convulsions when abroad. Paul Wride, aged thirty-nine, an epileptic from south London, was refused travel insurance but
decided nevertheless to join his friends for a dream holiday in
Canada. Whilst there he collapsed and was rushed to hospital in
Vancouver. The hospital refused to release Mr Wride before his
medical bills were paid. He was therefore stranded in Canada for
nine weeks and ran up a bill of about £97,000. If such insurance
blacklisting were to be applied to other conditions based on deleterious genetic variants and be adopted by other social agencies
such as immigration offices, employment exchanges, adoption
agencies and building societies, it is easy to see how a genetic
underclass could become established.

Government became involved when the Human Genetics
Advisory Commission requested a moratorium for two years to
prevent disclosure of any existing genetic tests to insurance companies until more information had become available on the
actuarial relevance of such tests. Another advisory committee to
the government – the Genetics and Insurance Committee – gave
the opposite advice. Genetic tests for a progressive dementia
(Huntington's disease) are sufficiently accurate and reliable to be
used by insurance companies when assessing application for policies and would reduce the problems of adverse selection. The
Association of British Insurers requested and was granted such

use by the government. The UK has therefore become the first country in the world to give official sanction for the use of results from genetic tests to evaluate premiums for life insurance. Ten other genetic tests for seven other diseases are awaiting similar approval.

Who are likely to be the winners and losers in all this? Since the insurance industry requested the use of genetic tests in the first place they may consider themselves as the winners. They hope to be able to avoid insuring bad risks at low premiums. But people buying insurance may become reluctant to have genetic tests in case they end up having to pay more. Alternatively, if genetic tests and disclosure become mandatory (as is family history) this could still spell bad news for the insurance industry. The genetically best off may decline to pool their risk with others, and the worst off will find that they have become uninsurable. Thus the insurance market may dwindle. The dilemma is between banning the use of genetic tests and perhaps destroying the insurance industry by adverse selection, and allowing their use and creating a genetic underclass who cannot afford to insure themselves. Insurers have as much to fear from the advent of genetic tests as do the purchasers.

In its document on Human Rights and Biomedicine (1997) the European Council at Strasbourg states that 'any form of discrimination against a person on the grounds of his or her genetic heritage is prohibited'. Even before the advent of genetic tests, and purely on the basis of family histories, there is already some evidence that insurance companies are dealing unfairly with relatives of families where a genetic disease occurs. The data come from a postal survey of 7,000 members of seven different British support groups for families with genetic disorders such as cystic fibrosis, Huntington's disease and muscular dystrophy. A comparison group of 1,033 subjects came from the general public. The response rate to the postal questionnaire was rather low, not unusual given the large number of questionnaires that are currently being circulated on so many different topics. About 13 per cent of respondents who represented no actuarial risk on genetic grounds, either being healthy carriers or being altogether unaffected, thought that they had been financially discriminated against on the basis of their family histories. They were asked to pay higher premiums or were refused insurance outright. This may simply have been due to confusion or ignorance about the genetic disease on the part of the medical officers employed by the insurance company, or possibly it was a way of the insurers trying to avoid adverse selection of people who were considered at increased risk of an inherited disease. Perhaps the respondents thought they were being victimised by their family history when in fact this was not the case.

Clearly the claims for actuarial fairness by the insurance industry and the claims for social justice by the European Council will require new forms of regulation to clarify this problem. In future the state may have to provide a safety net of some form to protect a genetic underclass. One possible solution, as touched upon earlier, might be to set up a type of state insurance scheme similar to the National Health Service, so as to avoid penalising individuals who have been dealt a bad genetic hand through no fault of their own. In this case compensation would be paid by the more fortunate members of society who through no merit of their own were dealt a good set of genes. Private insurance would still have a place alongside a state insurance scheme in the same way that private medicine operates alongside the NHS. To many people it may appear fair for the state to redistribute some resources for insurance purposes from the genetically lucky to the genetically 'underendowed'.

In the USA, where health insurance is commercially organised and based on a mutuality model, many people take out no health insurance at all. A great deal of recent legislation has been aimed at restricting the use of genetic tests by health insurance companies and, more broadly, of any 'genetic information' at all. Federal legislation enacted in 1996 forbids insurance companies from using even family history to exclude people from schemes for group health insurance. The State of New Jersey has gone further and bans the use of genetic information for any insurance or employment purposes without written consent of the person involved. They have not suggested any practical measures to prevent adverse selection but appear to be making it easier for some groups to obtain health insurance. This is to prevent the growing trend for health care in the USA being converted entirely into an economic commodity, sold in the marketplace and distributed on the basis that those who can afford to should pay for it, with no pretence of a social service.

Employment rights and genetic discrimination

Today at least six American corporations screen employees for sensitivity to toxic substances that they may encounter during their work and, reasonably, deny employment to any individuals who might be allergic. This has been extended to genetic screening. However, if employers have diagnostic information about the DNA of prospective employees they could choose only to employ people with the best possible health prospects and not bother to change or improve bad working conditions that lead to ill health. This would reduce costs and their compensation bills for future absenteeism and illness. If they provide health insurance as a business perk for the job, it would in the long run reduce their insurance claims and subsequently the premiums that they would have to pay to the insurance companies. For their part, rejected applicants would have to look for employment elsewhere, work which might be worse paid, further from home or just more unpleasant in some other way. At present the normal practice is to offer employment on condition of a satisfactory medical report. If this practice is maintained and genetic conditions precluding employment are specified in advance, some safeguards will exist. When a conflict of interest arises between institutions and individuals, the latter are usually on the losing side. Their employment interests need to be protected.

A clear-cut case of genetic screening resulting in denial of employment has already occurred in the USA for individuals who carry a single abnormal gene for sickle cell anaemia even though

this has no untoward effect on their health. They are called carriers of the disease. The sickle cell screening programme began
with the good intentions of reducing the birth rate of babies with
sickle cell disease, but eventually it turned into a form of public
screening and, because it was misapplied, it became a means of
genetic discrimination against one subgroup of the American
population. African–Americans were denied access to high–quality
employment and insurance.

   It worked like this. If both partners were discovered to carry the
sickle cell trait they were counselled to avoid having a child who
might be born with sickle cell anaemia. Laws for screening the
sickle cell mutation, often sponsored by African–American legislators, were enacted in seventeen states. In 1972 Congress passed
the National Sickle Cell Anaemia Control Act that provided for
research, screening, counselling and education of affected individuals. The preamble to this Act stated, incorrectly, that two
million Americans suffered from sickle cell disease. In fact two
million were carriers of only one copy of the deleterious gene that
gives rise to the harmless condition called sickle cell trait. Fewer
than a hundred thousand people who inherited two copies of the
defective gene had the actual disease. However, the American Air
Force Academy, acting on this erroneous statement, restricted the
entry of carriers to the Academy. Commercial airlines restricted
sickle cell carriers to ground employment only and African–
Americans found their career paths blocked. The justification
given for this was the fear that sickling of red blood cells in carriers
would occur at high altitudes with minor degrees of oxygen deficiency. There was no clinical evidence for this. Spokespersons for
African–Americans in the USA indicted the compulsory sickle
cell screening programme as a form of racial discrimination and
eventually the law was repealed.

   In another case a textile mill in the Danish town of Vejl allowed
226 of its workers in 1992 to be screened for a blood enzyme that
prevents them getting a severe lung disease. If, owing to a genetic
defect, they lack this enzyme (called alpha-1-antitrypsin) their
lungs become spongy and are liable to be attacked by severe

infections. The Danish trade unions pointed out that the company could have used the information as a reason for sacking those workers who lacked the enzyme instead of trying to control the textile-engendered dust in their factories. The Danish Parliament has subsequently debated a bill that could severely limit the use of genetic tests to prevent discrimination against employees on the basis of a genetic predisposition to disease. Employers may be able to apply for exemptions if they can show that the presence of the disease would endanger lives of others at work, such as piloting aircraft. But to deal with hundreds of possible exemptions could lead to a bureaucratic nightmare. Interestingly, the Danish Confederation of Trade Unions has opposed a blanket ban on genetic testing because it says it will hinder the detection of medical problems in the workforce. It claims it is not the technology that should be controlled but who has access to the results. People should be tested confidentially and the information should not be available to employers before a person is hired, or used to determine the conditions for company health insurance.

However, in some instances employers should have the right to information about genetic defects of prospective employees if this would put at risk the lives of themselves or others during the course of their work. Mutations of some genes can lead to certain forms of night blindness or colour blindness and clearly such individuals should not be employed as airline pilots or sailors, or be involved in driving vehicles at night. Similar considerations could apply to the genetic defects underlying narcolepsy, where affected individuals can fall asleep at a moment's notice. Many unexplained road accidents can probably be put down to this condition. In a job where there is exposure to factors causing cancer, such as that of a lifeguard, an employer may be acting irresponsibly if he fails to screen for genes that may influence the development of a dangerous form of skin cancer such as a malignant melanoma.

With the array of new genetic tests that is becoming available, new legislation will be required to balance the employment rights for individuals at risk with the employers' rights to have their work done competently and without possible risks that the employees may create for themselves or others in the workplace.

To address the problem of possible discrimination against the carriers of harmful genetic variants, they could perhaps be protected under laws related to Disability Acts. The UK Disability Discrimination Act of 1995 defines disability as 'a physical or mental impairment, which has substantial and long-term adverse effect on . . . the ability to carry out normal day-to-day activities'. Would this mean that it is unlawful to discriminate against an individual with a genetic predisposition, say, to a serious disease such as cancer or dementia that has yet to appear? Or would it be unlawful to dismiss a worker whose genetic predisposition might interact with environmental factors at work to produce a disease such as asthma? The knowledge of this predisposition might create sufficient stress and anxiety in the employee as to impair his work performance. Would this justify dismissal by the employers? Or should the Disability Discrimination Act be so worded as to offer protection? This would mean that discrimination against people with a genetic predisposition to a particular chronic disease would become illegal. As already mentioned, the Human Rights and Biomedicine Convention 1997 of the Council of Europe agrees with this and states that 'any form of discrimination against a person on the grounds of his or her genetic heritage is prohibited'. If the Biomedicine Convention is to be given any effect in English Law it seems likely that the Disability Discrimination Act will have to be amended to prohibit discrimination on the grounds of genetic constitution. A newer definition of disability, probably to be amended by case law, would therefore have to include the possession of genetic variants related to some diseases. The new technology creates a new category of individuals who may be without any medical symptoms but who are predicted to develop disease in the future, in other words, these individuals are at risk. At present there are relatively few but recent progress in identification of genetic markers will undoubtedly increase the range of conditions which can be detected before symptoms appear.

Other disability statutes suggest that it is unlawful to discriminate on the basis of future disability and an employer should not deny employment simply because a condition has been detected before it has actually begun to produce harmful effects. However, an employer's decision to terminate employment of an individual who has significant risk, for example, of a heart attack or epilepsy during usual working conditions appears reasonable provided that 'significant risk' is adequately defined and dealt with on a case-by-case study of each subject. Perhaps there should be no blanket rules or generalisations about a class of people who possess a particular genetic variant in view of the great diversity in the human genome and the presence of protective genes that make prediction all the more difficult. It can be more difficult to demonstrate such protective genetic interactions but they definitely occur. They can perhaps best be seen when the genetic variants are found that protect individuals from developing a disease. For example, there is a common genetic mutation of an enzyme that removes fat from the bloodstream occurring in about 15 to 20 per cent of the healthy population in the UK and USA. This mutation, however, is not deleterious and appears to protect people from developing high blood fats. In some studies it has been shown to protect them from early heart attacks. The same has been shown for some genetic variants protecting people from developing insulin-dependent diabetes. These protective variants will interact with the disease-provoking genes and produce a variable outcome for the person, making prediction of risks extremely difficult.

Two American presidents and pre-employment genetic tests

An interesting pre-employment issue was raised about the former US President, Ronald Reagan. It is now known that he has Alzheimer's dementia, and it is very probable that he had early stages of the disease when he was in the White House. Should elderly presidential candidates in the future be tested for a genetic predisposition to the dementias before they take office? They could make catastrophic errors of judgement before their doctors diagnosed the actual disease. Or would testing be an affront to their civil liberties? Your genetic code is part of your own individuality. You should be able to choose whom you share it with. You can agree to disclose your genetic code to your doctor if you want him or her to use it for your benefit. You perhaps can refuse to share it with your insurers; but should governments or employers ever have the right to dictate to you who should have access to your genetic code? It is a difficult issue and there are no obvious widely applicable rules about genetic privacy. It depends so much on individual circumstances and the possibility of injury to other people.

Former American President, Bill Clinton, made up his mind on the issue. He signed an executive order in 2000 that prevented Federal employers from requesting or requiring their employees to undergo genetic testing of any sort. It forbade discrimination based on any genetic tests that an employee might have taken in the past. It banned genetic classification of employees in a way that might deprive them of promotion, work overseas, or any other job opportunity. The new law also forbade Federal employers from disclosing genetic information to third parties. There were certain loopholes in the law. Testing for mutations that might be induced by exposure to radiation during work would be permitted, but no mention was made of doing genetic tests such as those for colour blindness, night blindness or narcolepsy that could impair a worker's performance, as discussed previously. There is pending legislation in Congress against genetic discrimination in the Health Insurance and Employment Act of 1999. If passed it would extend this protection to the private sector as well. As President Clinton said at the time: 'We must not allow advances in genetics to become the basis for discrimination against any individual or any one group.'

Crime

Could the discovery of a possible 'aggression' gene (described on page 242) be used as an excuse to reduce blame in law for the actions of, say, a violent mugger? In fact this was tried as a defence, in 1998, for Stephen Mobley who shot and killed a man during a robbery in Georgia, USA. His lawyers used the study to argue that Mobley had a genetic predisposition to aggressive behaviour and was not responsible for his actions. Their argument was not successful. The court maintained correctly that 'the theory of a genetic connection is not at the level of scientific acceptance that would justify its admission'. Mobley accordingly received the death sentence.

But there clearly are biological and genetic determinants of crime. One has only to think of the different frequencies of serious crimes committed by men or women, even when social backgrounds of poverty and emotional deprivation are taken into account. Of all the people convicted or cautioned for a criminal offence up to 1997 in the UK, 82 per cent were men. It is a surprising fact that one in four men in the UK will have a conviction for a serious offence by the age of thirty-one, whereas for women it is only about one in fourteen.

Twin studies have shown that if you have an identical twin brother who has been prosecuted for a criminal offence you have a 50 per cent chance of being convicted for something yourself, but if your twin brother is non-identical the risk goes down to about 20 per cent. (Admittedly some of the siblings of the non-identical twin pairs will be female, and therefore less liable to commit an offence in the first place.) Such evidence does not necessarily mean that genes are at work in these family studies. For example, some tattooed youngsters tend to have parents who are themselves tattooed, compared to children from a similar social class without tattooed parents. But there is no gene for tattoos. It is just a question of social attitudes and customs that tend to go together in families. But genes almost certainly do play a part in some aspects of criminal behaviour and the question now is not whether to accept the idea of an inborn tendency to crime, but how far it goes in explaining criminal behaviour.

This was recently tested in Georgia. A young woman, Glenda Sue Caldwell, was imprisoned in 1985 for killing her son in a fit of rage. In her defence she claimed that she was at risk of inheriting an early type of dementia (due to Huntington's disease), her father having died of the condition. The rage attack on her son was supposed to be the first symptom. The jury was unimpressed and she was sentenced to life imprisonment. A few years later in prison she started to show clear signs of Huntington's disease. The judge retried the case and admitted that the unfortunate woman was not to be held responsible for her actions. She was accordingly released. The legal system in Georgia accepted the notion that the effects of specific genes can render a person less responsible for their actions in a court of law. This occurred before the actual gene causing Huntington's disease was discovered. Now a test is available to tell for certain whether a person has inherited the defective gene.

English criminal law tends to limit the role of 'excuse'. Children, as minors, are excused. Some categories of mental disorder can lead to an excuse, but factors such as family background, social and economic hardship – all of which may provide an explanation for why someone commits a crime – are nevertheless out of bounds as excuses. Such mitigation would create problems of where to draw the line. It would tend to encourage crime if actions could be blamed on our genes.

Barry Kingston, a man with paedophilic tendencies (which could have a genetic basis if one extrapolates from other studies on sexuality), tried to get his conviction for indecent assault overturned on the grounds that his co-defendant had given him sedative drugs which made him unaware of what he was doing. This claim was initially found attractive by the Court of Appeal, but finally rejected by the House of Lords. In the words of Lord Mustill: 'To recognise a new defence of this type would be a bold step. The common-law defence of duress and necessity (if it exists) and the limited common-law defence of provocation are all very old . . . I suspect that the recognition of a new general defence at common law has not happened in modern times.' This would probably also be the reception accorded to a genetic 'excuse' for paedophilia or other types of biologically determined behaviour considered to be criminal. Genetic predisposition only yields information that is probable, and the relationship between predisposition and actual behaviour will always, to a degree, remain uncertain. So the new genetics is unlikely to undermine the traditional theories of justice. Many criminal lawyers believe individual responsibility to rest on the principle of capacity and fair opportunity to act otherwise. The argument that fair opportunity is compromised by genetic predisposition runs up against the difficulty that not everyone possessing the predisposition resorts to criminal behaviour. Actions have to be treated as intentional simply because a coherent social order demands it – if someone has a predisposition to criminal behaviour they can and should take steps to avoid the situations that tempt them into it. In the final analysis it is the law that judges people and not genetic or other scientific criteria.

9

Justice Without Imagination

Yet law abiding scholars write
Law is neither wrong or right
Law is only crimes
Punished by places and by times.

—W. H. Auden, 'Law Like Love'

Where could one look for authoritative and impartial guidance for the use of new eugenic procedures? An alternative source to religious and legal texts would be the various humanitarian declarations that have been recorded throughout the ages.

Humanist declarations

One of the earliest of these was produced by Hippocrates, the founding father of descriptive medicine, who wrote extensively on such subjects as wounds and ulcers, fractures, diets and disease prognosis. There are some statements in his Oath that are relevant to present-day eugenics: 'I will follow treatments which according to my ability and judgement I consider for the benefit of my patients and abstain from whatever is deleterious and harmful . . .' Apart from doing nothing that may prejudice the health of his patients, he further stated that professional confidentiality should be maintained and that doctors should not practise abortion. The Hippocratic Oath is still taken by some newly qualified doctors to this day.

Another landmark was the French Declaration of the Rights of Man in 1789. This arose from the long-standing despotism of the French monarchs combined with the arrogance of the French aristocracy and their oppression of the middle and peasant classes. This led in part to the French Revolution. The Declaration of Rights attempted to redress some of the injuries and injustices the population had suffered under the *ancien régime*. Two of the declarations are relevant to present-day eugenics. One is the autonomy and inviolability of the individual; the other that law should only prohibit actions that are harmful to society. Other statements that all men are born free and equal, and are equal before the law, have a general bearing. One could argue that cloned children would not be born equal to children born naturally, since the former may start with an inherent disadvantage from the many somatic mutations forced on them by the adult who donates the nucleus (see page 64 for further details).

The United Nations Charter on the Declaration of Human Rights in 1946 was drawn up in the wake of the multiple atrocities committed during World War II. It is a very wide-ranging document dealing with the rights to life, liberty and to protection against slavery. The Declaration is to be applied without regard to race, sex or religion. The Charter considers that it is everyone's birthright to marry and found a family of their own choice. Should this include choice of the gender for their children? Protection should also be provided for an individual's privacy, both personal and domestic. Regarding the ownership of another person, might a parent feel that he has more complete control over his or her cloned child than natural parents, since the parent provided virtually all the genes for its development? A provision is made that society should provide some form of health care and social support for its members.

Coming to the present day, in 1997 the Council of Europe produced a Convention on Human Rights and Biomedicine. This is so important that some of the basic details are summarised in the Appendix, Table 2. They echo UNESCO's Declaration on the Human Genome and Human Rights, adopted in November 1997, that 'no one shall be subjected to discrimination based on genetic characteristics that is intended to infringe . . . human rights, fundamental freedoms and human dignity'.

How much of this 1997 Convention will be incorporated into law? By 1999 five states (Denmark, Spain, Greece, Slovakia and Slovenia) had ratified the articles as a binding legal text for their countries. Twenty-two other states have signed up to the Convention since 1997, and many will proceed to ratification. The UK has claimed that some of the articles of the Convention will have the effect of inhibiting future research and so has not agreed to ratification. More recently the European Parliament in Strasbourg has published a Bill of Rights for 2000 and in Article 3 it calls for 'a ban on eugenic practices', obviously not quite understanding what the word 'eugenic' means. Banning all eugenic practices would mean prohibition for the screening for Down's syndrome babies or other severe genetic defects, and then vetoing a termination if the mother so wishes.

Many errors can lurk in these general statements and their interpretation can be difficult, but they all appear to agree on certain key features: respect for the individual over and above society, especially for his or her privacy and autonomy. A person should not be forced to act as a tool for society. Informed consent must be obtained and respected before any bodily intervention can take place.

However, people's ideas about respect for the dignity of individuals can vary considerably. A good illustration of this can be seen in the unlikely pastime of throwing dwarfs. In 1995 a French club devised a game to see how far their customers could throw a dwarf. The dwarfs fully consented to being thrown – provided there was a soft landing – and in fact the practice provided them with quite a lucrative source of income. The French Conseil d'Etat, however, ruled that dwarf-throwing compromised human dignity, and the game was banned. The law rescued the dwarfs from one situation that was deemed to violate their human dignity only to place them in another which perhaps even more seriously undermined their dignity – by making them unemployed.

A very serious example of failure to respect the dignity and rights of the individual was the Tuskegee study in Alabama, USA. Before the discovery of penicillin Swedish doctors had published an extensive account of what happens to men who become infected with syphilis. In 1932 the United States Public Health Service authorised a similar study with African-Americans to see if the disease followed the same course as it did in Swedish men. Accordingly, in Macon County, Alabama, 400 African-Americans between the ages of twenty-five and sixty with signs of early syphilis were recruited and systematically observed for the next forty years. They were never told the nature of the experiment and instead were given token medicines to keep up their interest in the study. After 1945, when penicillin became available for the treatment of syphilis, the subjects were never told that any effective treatment was forthcoming. However, the families of the men were given money to pay for coffins and burial as they died. By 1955 about one-third of the original number of men had died of syphilis of the brain or of the arteries. By the early 1970s, after many papers had been published in medical journals describing the course of the disease in African-Americans, it became a public health scandal. These men had never been told the nature of the experiment or given informed consent for it. The study was terminated in 1972 with an investigation before a medical council. None of the doctors, administrators or politicians who had been involved in the study was prosecuted; nor indeed did any of them admit publicly of wrongdoing. The African-American nurses who supervised much of the study were caught in the cruel middle ground between loyalty to the medical research project conducted by the white doctors and fellow feeling for the duped African-American patients. A few years later a civil rights lawsuit was filed against the United States Public Health Service but was settled out of court on payment of more than $9 million. No formal apology was ever made at the time and no admission of wrongdoing. Only twenty-five years later did President Clinton offer unreserved apologies on behalf of the US government to the surviving victims of this cruel and inhumane study. This episode illustrates how much ethical judgements can change over a short period of thirty years.

Legislation

Numerous factors will make the development of practical and effective legislation for the regulation of the new predictive genetics a very difficult task. These include: (1) the rapid pace of genetic discoveries with new technologies evolving annually; (2) a diversity of opinions on the appropriate application of these new technologies; (3) the paramount importance for scientists of preserving basic freedoms of research and communication if restrictions are to be imposed by statutory bodies; and (4) the evolving social norms of society regarding the use of the new eugenic techniques. All these will make it difficult to choose appropriate modes of regulation. These at present range from public referenda, to moratoria, to professional regulation by bodies such as the General Medical Council, to ethical codes to be agreed by government departments such as the Department of Health, to the creation of advisory agencies such as the Human Genetics Advisory Commission, and finally to the drafting of scientifically accurate and appropriate legislation.

A practical example of the problems of a legal approach was cited on page 146 when a two-year moratorium was requested in the UK during a conflict that arose between the Association of British Insurers and the Human Genetics Advisory Commission. The dispute concerned the use of predictive genetic tests for evaluating premiums for the purchase of insurance products. The moratorium was to give time for the insurance industry to collect data on the actuarial relevance of these predictive genetic tests. However, even if one were to take the simplest case of using Factor V Leiden as a predictive mutation for venous blood clotting, it is very unlikely that any useful actuarial data would accrue within two years. Carriers of such mutations may develop blood clots up to twenty years after the identification of the mutation, so long-term prospective trials would be needed to assess its predictive value for insurance purposes. The request for the moratorium was finally rejected by government on the grounds that the insurance industry should be allowed to run its own business without excessive interference by other agencies.

Although in the past justice has required regulation to act prospectively, in the present circumstances of such a fast-moving field a more flexible, responsive and retroactive regulatory model may be more appropriate than a rigid criminal or prohibitive approach as has been proposed by many countries. Prospective legislation is difficult to update given the rapid pace of genetic discoveries and appears to be very arbitrary. If you undertake experiments in the UK on an embryo older than fourteen days you can expect to be treated like a criminal and serve a prison sentence, whereas if the embryo is less than fourteen days old you are treated as a bona fide research scientist. No wonder such ruling is in danger of receiving such little public trust as to make it ineffective.

Another major reason why legal responses are going to be difficult is the problem of obtaining a consensus that is acceptable to all groups within our diverse and pluralistic society. As an example, the cultural diversity within two square miles of where I work in central London is enormous. In 1997 there were eight different ethnic groups. They comprised: Caucasians (57 per cent); Caribbeans (7.9 per cent); Africans (6.7 per cent); Indians (6.5 per cent); Pakistanis (3.7 per cent); Bangladeshis (10.2 per cent); and Chinese with other Asians (3 per cent). There are as many different social norms as there are religious groupings and a recommended procedure that is acceptable to one group may be highly offensive to another.

Christians, Muslims, Hindus, Buddhists and Confucians all have differing beliefs, so it was surprising to find that they share a number of common beliefs when asked to complete an elaborate questionnaire for their views on fundamental human rights. Although the different religious groups appeared to agree about common rights, no one could explain why. Perhaps they were similar simply by virtue of being human, regardless of what religion they follow. There has of course been a greater trend to globalisation of opinions in recent years. If the survey were to have been conducted a hundred years ago many more differences may have been apparent.

To date there has been very little consistent legislation with regard to predictive genetic testing. The major legislation has been in the field of pre-implantation genetic diagnosis where it is illegal to use gender testing in the UK (but not in the USA) or any other tests, unless it be for the prevention of a sex-linked disease such as Duchenne's muscular dystrophy or other specified forms of severe inherited disease. It is also illegal to undertake embryo splitting in the UK (but not in other EU countries such as Italy). This is because it involves manipulation of embryos that is regulated by the UK Human Embryology Act. These statutes are likely to come under direct challenge now that the new European Convention on Human Rights Act has been incorporated as a Bill of Rights in UK law.

The legal issues regarding informed consent from under-age children who want a genetic or any other medical test for a disease illustrate other problems. One of my patients, a twenty-one-year-old who saw his father develop an early and genetic form of Alzheimer's dementia, wanted to know whether he would be likely to develop it too. The test was ordered with no problems. However, the Family Law Reform Act of 1969 states that if he had been fifteen years old he would not have been able to give proper informed consent for the test until he was older than sixteen. Even if he had a higher intelligence or a better understanding of the condition than his mother it would still be unlawful to perform the test. Of course, his mother or anyone else with parental responsibility could have given consent on his behalf. The arguments for withholding the test, especially for an as yet untreatable disease such as Alzheimer's dementia, are that adolescents are particularly vulnerable; although the young man in this case understood the nature of the disease, having seen his father undergo an early dementia, he might not at an earlier age have been able to cope emotionally and socially with the consequences if the results indicated a bad prognosis.

However, if the young person's values and identity seem reasonably coherent and secure perhaps there is less need for such a paternalistic law. It all depends on the child's best interests and who should determine them. If he were considering taking up an occupation that involves the possibility of repetitive head injuries, such as professional football, perhaps it would be very important to know as soon as possible if he carried genes that predispose to Alzheimer's dementia. In a way genetic tests merely exchange the uncertainty of *whether* the dementia of Alzheimer's disease may develop, for the uncertainty of *when* it will develop. In either case there is the burden of uncertainty for the adolescent. The question therefore becomes not whether the adolescent is competent to consent to the test but whether the potential burdens of the test results outweigh the benefits. Clearly if the test is for a treatable or preventable disease, predictive testing in childhood can be beneficial. When the test is for an as yet untreatable disease, such as Alzheimer's dementia, then perhaps each individual case should be taken on its own merits and the law should not be asked to make general pronouncements about age of consent for this particular issue.

The problem, as always, appears to be where to draw the fine line between an individual's basic rights and interference or control by society. If an infertile couple want a child and all the other techniques have failed, has society the right to forbid them by law to use cloning as a last resort? The law would claim that by banning cloning it is protecting the rights of the unborn child. But of course one cannot know if the child is going to be damaged or harmed by the technique until it is born or even grown up.

The law does, however, interfere reasonably with some of our most private and intimate relationships. Sexual relations with a youngster before the age of consent is a criminal offence. Since 1997, in England and Wales, if you know you suffer from HIV infection (*H*uman *I*mmunodeficiency *V*irus) and have intercourse without warning your partner, this constitutes a criminal offence. In 2001, Stephen Kelly, aged thirty-three, was the first man in Britain to be convicted for having sex with his girlfriend without telling her that he was infected with HIV. But against this were the unreasonable laws against homosexuality (and not lesbianism) in the nineteenth and early twentieth centuries. These are now seen to have been discriminatory, unjust and brutal, producing untold misery and anguish for the men involved. (Oscar Wilde's *The Ballad of Reading Gaol* expresses this eloquently.) Law perhaps should move away from the old paternalistic approach of proscribing public behaviour and move towards collaborating and negotiating with the public more as partners in the law-making process.

Negotiations with the public

Public meetings, television debates and consultations with all parties concerned can be used to clarify major issues involving the new eugenics. All the different viewpoints and opinions of the academics, applied scientists, industrialists, government and the public can be aired and when possible harmonised, despite the widely differing agendas. Academics want the freedom to pursue research and follow wherever their most interesting and exciting observations lead. Freedom for research is as important to them as freedom of speech. They are fiercely competitive for scientific reputation and research grants to develop their work and often like to establish large departments with themselves at the head. Increasingly they are interested in forging links with industry. Applied scientists on the other hand are often under some sort of control by a board of directors of a company; they are not usually free to do research according to their own interests. There must be potential applications of their work that are feasible, safe and eventually profitable. The primary concerns of industrialists are whether the products of their research are safe and commercially viable. Government is in a more complex position. Receiving party funds from industrialists and at the same time relying on public votes to keep them in power requires a delicate balancing act. If they cause damage to the public's health (as in the mad cow disease debacle) or bow to the profit motive of multinational companies (as in the recent debate on genetically modified foods), they risk losing their power base. Finally, the public want products that make their lives easier and happier; they want health for their children and themselves.

How these conflicting aims can be partially harmonised can be studied in one special area of public concern, namely the health care of patients. The players here are patients, doctors, nurses, paramedics, medical research scientists, the drug industries and government agencies. In the past patients were given advice and sometimes forced to do things that the doctors considered in their best interests. They were clinically examined, tested and underwent operations under very paternalistic conditions. I remember asking a patient with a large abdominal scar what had been done, and he replied that he did not know but had left it to the doctors who knew best. The public mood has swung considerably in thirty years. The public now want to be treated as partners in the therapeutic process. The overriding demand now is for information about the health problem, for giving informed consent for any medical intervention, for confidentiality about the disease and treatments, for minimising the risks of possible harm arising from the treatment and for maximising the possible benefits of the final outcome. Checks and controls are in place in the form of local hospital ethical committees. At a national level the work of the General Medical Council ensures that doctors perform within generally accepted professional standards. Recently the General Medical Council has come under attack as being more of a professional clique designed to protect the interests of doctors rather than the welfare of patients; the infamous serial killer Dr Harold Shipman, and the unauthorised retention of body parts by Alder Hey Hospital in Liverpool, have added fuel to this. Loss of public trust in its work has led to the likely future imposition of external audit and controls on its activity.

The general public, as consumers of eugenic products, should be given similar types of safeguards. In addition they face the pressures and demands of a free marketplace. Aggressive advertising through all the media channels evokes feelings of unfulfilled desires and needs, or instils fears that can only be satisfied by purchase of a particular product in our brand name society. Might you have an early bowel cancer? Then why not buy our home test to check this out? Worried about the paternity of your child? Contact us at www.dnanow.com: from a sample of four hairs from you and the child we can tell with 99 per cent accuracy whether you are the father or not. Are you unable to have a child? Let our company produce one for you by cloning. Do you want your child to have specific characteristics? Let us design one for you on the Internet (go to www.genochoice.com). This approach opens the door to the idea of consumer eugenics with parents considering what can be done to ensure the 'quality' of their future children. There will be no way of avoiding this type of eugenics in the future. A major task will be to determine the limits beyond which we should not go. Eugenics may be acceptable now to prevent disease; but how far should it be used to reflect our choice of social values for our children that relate to intelligence, beauty, athleticism, etc.?

Public attitudes to the new eugenics appear to be informed mainly by the leisure media where one soon discovers that the latest genetic news stems not necessarily from scientific facts but more from lobby groups and spin-doctors. Sunday supplements report 'exclusive' features with overblown stories. Television programmes focus on the helpless grief of families where the genetics have gone horribly wrong. The cinema has coloured the public's imagination more than any of the popular science books. The general public has easy access to the media but very poor access to the committee tables of the regulatory bodies where the hard decisions are being made on how to regulate the new technologies. The committees' reports often deploy a great deal of knowledge and intelligence but without too much human interest. Somehow democracy has to become more accessible. Politics should not only crop up at election time. People need to have a more genuine, informed and continuous say on the matters that directly concern them.

To redress this imbalance consultative procedures such as citizens' juries, deliberative polls and public survey of attitudes have been developed. Deliberative polls occur when several hundred members of the public are recruited at random to survey their views on a particular issue by means of a questionnaire. After this they convene for a weekend to debate the issues, and the questionnaire is then repeated to detect any changes in attitudes after a full discussion.

One of the more unpleasant consequences of the poor access of the public to the decision-making process has been the formation of aggressive pressure groups and the action they have taken. Dr Barnett Slepian of Amhurst, New York, was shot dead through his kitchen window by a sniper working for an anti-abortion group in the USA; an unknown gunman fired shots into the children's playroom of the house of Dr Robert Crist who was on a 'wanted' list by anti-abortion protesters. Since 1993 seven doctors who performed abortions in the USA have been murdered, and there have been fourteen attempted murders by 'pro-life' pressure groups. Certainty is their stock in trade. During the last two decades more than 2,300 incidents of violence against abortion clinics, such as arson and bombing, have been reported. The anti-abortion lobby, further blinded by invective and inflamed by ugly passions, has established a website where a 'hit list' of doctors called baby butchers can be found in the 'Nuremberg' files. These files are decorated with embryos dripping blood and a line is drawn through the names of those doctors on the list who have died. Stung by these attacks, opinions have polarised and an opposing pressure group has been formed called Planned Parenthood. A group of 'pro-choice' doctors have brought a civil action against the anti-abortion groups, contending that their website constitutes incitement to violence and intimidation against them.

Everyone of course has the right to deliver their opinion, but with the appropriate degree of moderation and courtesy. There needs to be more mutual respect for each other's viewpoints. Opposing interests should be considered, must be compared and if possible be reconciled before proceeding to the courts. The 'pro-choice' group were provoked well beyond the necessity of the case. They went to court and were eventually awarded $107.6

million in damages by a jury of four men and four women in Portland, Oregon.

The passion and commotion of public attitudes to abortion can vary enormously. At one extreme some people consider abortion at any stage from the time of conception to be a form of murder, whereas others want far better reasons for having children than not knowing how to avoid having them. Their belief is that we need a world in which fewer children are born and in which we take better care of them.

Genetically modified food

The power and sway of public opinion is no better exemplified than in the dispute over the use of genetically modified foods in Europe. The UK government, in collaboration with biotechnology companies such as Monsanto, initially gave the 'go-ahead' to introduce genetically modified foods on to the shelves of our supermarkets, but public trust and confidence were not won. Public demonstrations against genetically modified foods caused them to be rejected by the large UK food retailers such as Sainsbury's, Tesco and Asda and the consumer backlash against genetically modified foods reduced the demand, forcing the profitability and share prices of biotechnology companies such as Monsanto to fall sharply. Some smaller agro-technology companies even went bankrupt.

This anti-technology campaign in Europe and then the USA was not based on any reproducible scientific evidence or any real philosophy for its support. It was just a gut feeling that genetically modified foods were wrong – a sort of People's Inquisition. In May 2000, Lord Melchett and his twenty-seven colleagues were acquitted of trespassing and causing criminal damage to a farmer's crop of genetically modified maize. They were greeted as heroes at Norwich Crown Court after the jury of six men and six women were unable to reach a verdict on criminal damage. The defence argued that Lord Melchett and colleagues damaged the farmer's property to prevent damage to other crops in neighbouring fields by contaminated pollen. There is no evidence that any harm would have ensued. It appears to give a *carte blanche* to other forms of trespass and vandalism on the basis of unconfirmed fears.

Such actions have defined the limits to which big corporations may aspire and have had a baleful effect on Third World agriculture. Some parts of the developing world such as sub-Saharan Africa and South America desperately need genetically modified crops to improve their own food supply and prevent widespread malnutrition. India was hoping to combat the leading cause of blindness in children by growing a form of genetically modified rice that contains large amounts of the precursor of vitamin A, the nutrient deficiency responsible for loss of sight. Such countries naturally feel aggrieved that well-fed and wealthy Westerners have forced an embargo on the experimental development of genetically modified crops by biotechnology companies. Although there may well be hazards in the use of genetically modified crops, we at least need to do the proper experiments to confirm or disprove them.

Admittedly some technologies are dangerous but one could not live without them. Electricity can kill you; but it also runs our lighting, heating, our trains, planes and computers. Michael Faraday (1791–1867), the discoverer of the uses of electricity, was asked by the politician William Gladstone, 'What is the use of electricity?' To which it is reputed that Faraday gave two answers: 'Of what use is a new-born baby?' And then, after thinking some more, 'Well one day, sir, you may tax it.'

On the other hand there are too many examples in the past of technological advances that have gone wrong. Asbestos is a great insulator, but just happens to cause lung cancer if the dust is inhaled. Chlorofluorocarbons (CFCs) are great refrigerator coolants but just happen to destroy the ozone layer. Lead added to petrol certainly improves its efficiency as a fuel but just happens to pump poisonous lead fumes into the atmosphere that can cause dementias. Oil-based fuels are cheap and efficient but just happen to emit carbon dioxide and other greenhouse gases that are affecting our climate to cause global warming. It looks like a case of et cetera, et cetera, as we 'advance' on to the next untested technology.

Future issues

Bringing an acceptable framework of law and order to such a medley of different and hostile opinions for the newer techniques of embryo screening or cloning is not going to be an easy task. It is going to be difficult to make laws that correspond to what a majority of people will believe to be just, especially when the complex issues and varied opinions about the new eugenics do not make for easy navigation. Some people will prefer sticking to what is legal even if it does not match their most closely held beliefs. Others perhaps will value their principles more highly than statutes perceived to be unjust. It has been said that if you cannot abide one or two unjust laws then you probably do not deserve a perfect set. A bad law may be a suitable subject for censure without becoming an object for repeal if this were to bring down a whole set of interrelated and beneficial rules. But the real weakness of even our best laws related to the family and eugenics is that they cannot be proved to be of benefit for every particular situation to arise in the future, as we have seen with the case of Diane Blood.

It seems clear that a more flexible, responsive regulatory model would be more appropriate. New concepts and techniques could be allowed to develop freely and, if proved safe, could be put into practice. Retroactive rules and regulations could be devised once the outcomes are known. A multifaceted approach involving professional and self-regulation, moratoria, referenda and amendments to human rights legislation should be able to accommodate the rapid pace of genetic discoveries and techniques. The approach should also be able to respond sensitively to the diversity and changing social norms of the public as new knowledge is gained.

Justice

For social justice to be done at least two major issues have to be satisfied. Firstly, the regulatory framework should primarily be made for the benefit of the community but at the same time without injustice to a single individual. The regulation has to be equitable and not transgress the common rights of citizens. Fundamental requirements are: (1) respect for individual privacy and autonomy; (2) personal reproductive choices to be uninfluenced by external pressures related to genetic discrimination; (3) informed consent required for any interventions; and (4) rights to as 'normal' a family life as possible.

The second issue relates to a form of distributive justice: who should get what, and who should give it? How much of our scarce financial resources should be diverted away from areas such as resuscitation of very premature babies who may never develop properly into healthy adults, or from the resuscitation of very elderly patients to give them a few more years of a very restricted life? How much should be allocated instead to the new eugenic techniques?

Allocation of resources is a form of medical rationing. We cannot do everything that has become possible simply because the costs would be enormous. Some form of pragmatism has to be adopted. Should the new eugenics be left to the forces of a marketplace, where those who can afford to pay will have access to the services and the rest of the population may be left in a genetic backwater, thus fostering the ever-widening gap between the haves and have-nots in society? Already in the UK about 80 per cent of the new reproductive techniques are being carried out privately and only about 20 per cent within the National Health Service.

Equality, often called the mother of justice, should ensure the same access to the new technologies as for all other medical therapies. The new technologies should be for the benefit of everyone. If not, over a long period of time unequal access to the new genetic techniques could even split *Homo sapiens* into two separate subspecies. This is not as naive or far-fetched as it seems. There are already two examples in the line of early man's evolution. Man's ancestors who lived in Africa about three million years ago, the Australopithecines, evolved into two separate groups probably on the basis of the different methods they employed to secure their food supply. During the separation, climatic changes possibly converted wet forests in South Africa into a drier scrubland necessitating a change in feeding habits. *Australopithecus robustus* became more vegetarian in their food habits, as deduced from the structure of their jaws and dentition making them suitable for chewing plant materials, whereas *Australopithecus gracilis* became more carnivorous, employing stone tools to get meat. So in some ways differences in technology involved in food gathering may have helped to separate the precursor Australopithecines into two distinct groups.

Another example is the two probably distinct species of humans that roamed the hills and valleys of Europe about 30,000 years ago. Modern man, *Homo sapiens*, can easily be distinguished from the Neanderthals by differences in the fossil bone structures of their skulls and limbs, by differences in the number and variety of flint tools that they made, and by the bodily ornaments they manufactured out of shell and bone. But the Neanderthal fossil record suddenly disappeared about 30,000 years ago. One theory has it that there were climatic changes in Europe at this time and *Homo sapiens*, being more adaptive and inventive, survived by adopting a different lifestyle to suit the changing conditions. The Neanderthals had a less developed tool technology and became extinct because they could not invent new solutions.

Other possible scenarios propose interbreeding, for which there is slight evidence; or the destruction of the Neanderthals by *Homo sapiens* (which, given our present-day behaviour, seems much more likely). So modern man took over the whole planet, eliminating all competing hominid species? The point is that two types of humans evolved in separate directions because one was able to invent novel and superior technologies.

This could happen again in the future if there were to be a large enough disparity in technology. Perhaps *Homo sapiens* will eventually be replaced by a new species of man, *Homo technicalis*. One can see early signs of this happening already, for example, in the differential access to the drug technologies needed to treat HIV infection in young Europeans compared to young Africans. The United Nations joint programme on HIV infection forecasts that up to half the teenagers in some parts of Africa will die of

AIDS. In highly infected areas such as Botswana, two-thirds of all fifteen-year-olds are predicted to die of the disease before they are fifty. This could drastically alter the demographic balance of the white and black races for the foreseeable future.

However, equality and freedom of access to even the least costly of resources can be difficult to secure. Half the population, i.e. women, were denied the vote in America until they won suffrage in 1919–20; and in Britain until 1928, when the right to vote was extended to women under thirty. In France it took until 1944, in Canada until 1950 and in Switzerland until 1971. It seems that the words 'equality' and 'freedom of access' are easy to say but putting them into practice is a much harder job.

10

From 1984 to 2084

'From the age of uniformity, from the age of Big Brother, from
the age of Doublethink, greetings.' . . . Winston woke up with
the word 'Shakespeare' on his lips.'

—George Orwell, *Nineteen Eighty-Four*

Two revolutions are changing our lives at a rapid pace. One, driven
by computers, is giving us quick and easy access to an enormous
range of information via the Internet and a radically different
marketplace. The other, driven by the knowledge and power of the
genome, can give us designer babies, cloning and the prediction of
life expectancy or occurrence of serious disease. The two technologies have merged to form a new subject called bioinformatics.
This is required to work out the significance of the vast amounts
of data generated by studies of the genome as well as other biological fields such as the nervous system.

However, the more that science and technology intrude into
our society, the more they can produce a feeling of dehumanisation
and alienation. New technologies can transform people's lives and
create new social classes. By developing factories for spinning and
weaving cotton, the Industrial Revolution of the late eighteenth
century in England created a large new social class – the proletariat. The technology of the new eugenics may have similar
powers of class division. It seems that it is already outrunning its

sociology. Society then becomes a simple reproductive mechanism that has to be oiled and greased to be kept in good working order by the eugenic specialists and other experts. This again could create a new social class – a genetic underclass.

The applications of science often reinforce social inequalities instead of remedying them. The poor, squatting twelve to a room in the slums of Lower East Side, New York, or in Rio's *favelas*, are barely tolerated and live their rickety lives on the margins of our society. They are bewildered and unable to do anything with the new eugenic or information technology. How can it help them with their daily problems of getting enough to eat, finding enough money for the rent or to make down payments, back payments or pay other daily bills? No social workers alone could possibly improve their conditions.

The rest of us, leading lives of relative luxury, are challenged (and sometimes dazed) by having so many new things to learn. There are unparalleled demands on us to make choices about issues we do not properly understand. Different experts give conflicting advice, adding to our state of uncertainty. After much research and erudition on our part we resort to vague and vaguely elevating phrases such as 'faith in science', 'scientific progress' and 'power to control the biological world' to sustain our endeavours. We somehow hope that the syntax will help us to make some sense of all this. Some of us are still duped into believing that science is omnipotent, whilst others feel that science and technology are not the most appropriate ways to advance our society. But it is laughable to think of progress in terms of being led no one knows where.

Rather than continuing to accumulate wealth, learning or even new knowledge, our society needs to evolve a measured, considered set of values and ethics concerning the use of the new biotechnologies that relate to eugenics. We need to develop a definite attitude to the past which some of us do not even have the courage to face when it comes to the history of eugenics. Some people deny the Holocaust. We seek excuses in the aberrant behaviour of dictators, ignoring the mass of the population that was swept along by their ideas.

Yet the ethics of the new eugenics is the one that interests some people above all others, in some ways even more than the actual science that has produced the problems. One can become weary of scientific analysis, dealing continually with minutiae of experiments and the small predictable truths that often emerge. (Such minutiae are rarely of vital importance.) However, there are so many ethical and moral questions arising from the applications of the science and so few really satisfactory answers. Looking ahead one senses a crisis. The way looks full of intricate and perplexing mazes. The peace, stability and future health of the next generation could be at stake in every decision we now take. One wonders which is the worse path – to keep to the status quo of near zero-growth development, or to rush ahead with the new untested technologies.

The values we adopt will have an influence on all the actions and reactions that follow. They will excite and provoke all the passions and emotions attached to some of our most deeply held beliefs. But a coherent set of ethical principles governing eugenics is going to be very difficult to establish prospectively to cover all possible cases. No sooner has one set of rules been established than the methods and techniques change so that the rules need to be modified retrospectively. We are often left with just a medley of opinions and beliefs to justify our course of action.

As we have already seen, many new technologies considered outrageous at the start became so familiar that they no longer give rise to dispute. This gives way to the next set of pioneering breakthroughs that are developed on the foundation of the older technology and generate new sets of dilemmas.

In the 1940s two women doctors, Barton and Jackson, developed the novel technique of the artificial insemination of women from infertile marriages with semen donated by anonymous men (*A*rtificial *I*nsemination by *D*onor, or AID). The doctors were sympathetic towards childless couples, and believed that they should help women who were condemned to sterility through no fault of their own. There was an immediate and hostile outcry against the procedure. The Church condemned the technique as a vicarious form of adultery; and indeed such a case came up before the courts. It concerned a woman who had been to the USA for artificial insemination and had become pregnant. On returning to England she was sued for divorce by her husband on the grounds of adultery.

The technique of artificial insemination was considered to be a frontal assault on the sanctity of marriage and family life. Children were being created by a mechanical trick. It was suggested that the mother would continually dwell on the phantom father to the exclusion of the social father. The child would grow up to spend fruitless years searching for the 'real' father, and so on. In 1945 many members of the medical profession roundly condemned the procedure in letters to the *British Medical Journal*; they put forward all sorts of theories: 'the child would be illegitimate and have no rights in law', 'the child would learn he was a result of a planned breeding experiment and this would eventually lead to forming human stud farms', 'that it encourages masturbation which is a sexual aberration akin to sodomy', 'that the procedure is illegal, immoral and disgusting, and the most diabolical thing ever heard of; the whole procedure reeks of vicious expediency, half-truth and advocacy of plain lying; and would assuredly cause a break-up of Western civilisation'. Quaint opinions when read today. Nonetheless, opposition became so great that a Royal Commission was set up to see whether artificial insemination should be made a criminal offence. It was not made illegal, but neither were attempts made to regulate it since this would imply tacit approval of the procedure. Fifty years later the technique is in such widespread use that it goes almost unnoticed. We forget that the intellectual dogmas representing current opinions on scientific progress are never final. They are no more than just another storey added to the Tower of Babel.

Ethics and the new eugenics

Some people's ethical beliefs (including my own) are rather like icebergs. About one-seventh are clear and well formulated whereas the other six-sevenths are obscure and submerged below murky water. What resides below the surface may be the most important when a titanic problem appears on the horizon, and can lead to disaster.

What now follows is a view of ethics adopting the lines of reasoning from utilitarianism. From this point of view one's values are more a matter of statistics than religious texts: 'The greatest happiness of the greatest number is the foundation of morals', as promulgated by the philosopher Jeremy Bentham (1748–1832).

With each new scientific advance, and the technology developing from it, the problem arises as to the rules to adopt for their application. The arrival of new subject matter sometimes has to change our ethics because old ethical solutions to new problems may prove quite inadequate. Knee-jerk responses such as the US Senate's attempt to make human cloning a criminal offence were not particularly helpful. The Senate narrowly missed criminalising the procedure by six votes. If the bill had been passed its effect would have been more to encourage the formation of a cadre of outlaw scientists working in other countries to satisfy the real, although probably small, need of a few wealthy individuals who want a genetically related child by any means.

As it is, several American scientists such as Dr Richard Seed (as discussed on pages 66–7) are claiming that they will clone a child within the next few years. It is not illegal in the USA but most medical scientists consider this unethical and unwise at our present state of knowledge. If it were to be made illegal, as in many European countries, this would probably have the unfortunate consequence of driving the procedure underground.

Arguments from authority
A major problem is that there is no authoritative analysis and guidance as to how these new eugenic techniques should be used. If the Deity had foreknowledge of the new genetic developments to come He/She certainly did not reveal any attitudes to them in the extant scriptural records that are attributed to Him/Her. (No

self-respecting modern woman would subserve a patriarchal Church that oppresses women.)

No eugenic formulations are laid down in the higher authorities such as the Bible, the Koran or other scriptures. None deals with the present-day intricacies of genetic markers, embryology or gene transfer techniques. For abortion or euthanasia, however, as opposed to eugenics, there are explicit biblical injunctions. It is commanded: 'You shall not commit murder' (the *New English Bible*); the Koran takes the same line. Murder is wrong in itself, regardless of any consequences. But the Old Testament may not be entirely appropriate for this particular commandment nowadays. The Netherlands, a Protestant country, has passed a new law decreeing that voluntary euthanasia no longer counts as murder; it is now lawful. Previously, any form of euthanasia had been considered a criminal offence but there was nonetheless little prospect of prosecution for those who carried it out. It is believed that more than 50,000 cases of euthanasia have occurred illegally in the Netherlands since 1984. The last official figures from 1995 show 3,200 euthanasia cases for that year and another 400 doctor-assisted suicides. The new Dutch bill will now guarantee doctors immunity from prosecution provided that they adhere to a set of strict criteria. These include the patients making informed and repeated requests for termination of life, and that they must indeed be suffering from a terminal illness. A second opinion must also be obtained from an independent doctor who has examined the patient and agrees that the individual's life should be ended. The doctors must give the details in writing to a review committee after each case. If they breach any of these conditions they can be prosecuted for a serious criminal offence. The proposals perhaps give too much freedom to the doctors since the judgement is based solely on their reports and no lay people are involved.

The Dutch were not the first to introduce euthanasia. In July 1996 the Northern Territory of Australia enacted a law to legalise voluntary euthanasia and physician–assisted suicide, but it was repealed two years later. In the USA only the State of Oregon has legalised euthanasia. Its Death with Dignity Act (1997) stipulates that the patient must be terminally ill and must administer the lethal dose himself. Such laws could be copied elsewhere.

In 2000, a Court of Law in England delivered judgement in the case of Siamese twins Mary and Jodie. It ruled that the killing of one twin by surgery needed for their separation was not considered to be murder if it saved the life of the other twin, even when the operation was against the wishes of the parents. Jodie, the surviving twin, has recovered remarkably well after surgery and has vindicated the doctors' determination to go ahead with the separation, despite widespread condemnation.

This is the first time in the UK that surgery or any other medical intervention has been legally sanctioned against the interests of a patient (the weaker twin) and that inevitably resulted in her death (judicial murder). It is another example of the weakness of law that cannot foresee every possible contingency, even in such a clear-cut case as the murder of a defenceless child. How much more difficult are the new eugenic dilemmas going to prove for legal judgements.

The issue for voluntary eugenics appears to become much more a matter for contemporary opinion and beliefs to justify our decisions. Since there is no clear scriptural guidance on the rights or wrongs of the case, the responsibility of moral choice seems to rest squarely with us. The timescale is too short to be able to evaluate the long-term effects of any of the new techniques. We cannot form our ethical judgements on the basis of the consequences of the use of the techniques. They will not be known for at least one or more generations. The new eugenics has and will continue to pose difficult ethical problems rather similar to the current disputes about abortion. As our knowledge advances, disputes will arise on the use of selected donors for sperm or eggs, on the use of designer babies, the use of embryo enhancement, the cloning of humans and the prediction of chronic disease or longevity.

To explore one such dispute, reconsider in greater detail the shrewd and elusive arguments about abortion. These provide lessons in the way that future issues might be resolved when it comes to the adoption of the new eugenic technology. The dispute is caused by two different groups in society holding passionately opposed views. One argues convincingly that abortion is a crime akin to murder. They quote the biblical Commandment in support. The argument then revolves around the issue 'when does a human life begin?'. Is it at the time of fusion of the sperm and egg? Or is it at a later stage when the bodily organs of the embryo start to form? If the former, then the view that 'abortion is murder' is correct. If the latter, then the anti-abortionists appear to be over-reacting.

The anti-abortionists can point not only to the scriptures but also to the argument from rational ethics to support their own view. As usual, the Ancient Greeks came first to this subject, notably Plato and Aristotle. Aristotle's extraordinary book, the *Nichomachean Ethics*, attempts to find a rational basis for our rules of conduct, even to the point of using geometrical diagrams to clarify the problems, preferring not simply to leave the rules to be determined by custom or belief. This was taken up by Immanuel Kant, who tried to derive fundamental rules of conduct based on reason alone and not just in obedience to a higher religious authority. He attempted to discover the highest good, or *summum bonum*, as the foundation of all morality. His first universal rule, the first 'categorical imperative', is 'to judge your actions as if they were to be adopted as a law by all rational persons'. If doing so, would society continue to function properly? It works well for most of the Ten Commandments. If everyone by law were made to steal, murder, perjure themselves, commit adultery, disregard their parents, etc., social life would become intolerable. The same applies to abortion. If everyone did it where would the next generation come from?

But there are many loopholes in Kant's universal rule. Consider the ancient eugenic practice of exposing deformed babies overnight to see if they survived. If everyone did this, society would still continue to function as usual. However, most rational people would consider the exposure of infants to be an inhumane practice and it has been made illegal by most civilised countries. Yet another procedure such as abortion is still legal.

The pro-choice groups use arguments about abortion that are equally convincing. First: it is surely the mother's right to use her body in whatever way she thinks fit with regard to childbearing. No one has the right to impose their views upon her. Society should protect the mother's rights. Second: it is irrational and inconsistent to give a ball of cells the size of a small pin head (see Figure 2, page 24) the status and rights of a human being. It cannot in any way survive outside the mother. It has of course the potential to become a human being, but so has every egg that the woman discards before she menstruates. Third: should she then take steps to ensure that every egg she produces is taken through its development into a child? Surely not – depending on her particular circumstances she should be allowed to decide for herself whether the egg's development should continue or not.

The opinions of opponents and proponents stand in stark contrast. Who can decide between them? Justice in this case appears to be two-faced. One needs perhaps a sophisticated understanding to accommodate the views of both. One view considers what is just for the individual; the other looks to a fundamental principle of not destroying life. Each argument from its own point of view is incontrovertible. Any choice between them on the grounds of justice appears to be arbitrary depending on definitions of when a human life begins.

It then becomes a matter of analysing the consequences of the action for the people concerned. For anti-abortionists only their morals and principles are affected, whereas the woman's health or wellbeing may be harmed if she has an unwanted child. The anti-abortionists are in themselves not directly harming the woman but perhaps they are doing something almost as bad. With their mania for setting people to rights according to their own moral code, they restrict the freedom of action of the woman to pursue her own ends and welfare. The utilitarian dictum of the philosophy of Jeremy Bentham may apply here: 'Everybody to count for one, nobody for more than one.'

The contradictory opinions about abortion are even being reflected in UK law. In England a woman can have an abortion for medical reasons freely on the National Health Service, but in Northern Ireland, abortion is still a criminal offence (unless a woman's life is in danger). There a woman risks imprisonment if she has one. She has to travel to England if she needs to terminate her pregnancy for whatever reason. More than 6,000 Irish women (from the North and the Republic) did just that in 1999. Does legal consistency matter? Such an absurd legal situation could be resolved by simply allowing people, on a case-by-case basis, to make their own ideological choices in the matter, for or against termination. Let us hope such a solution will be adopted for many of the new eugenic techniques looming on the horizon.

Arguments from ethical relativism

If we cannot plunder any of the scriptures or sacred texts for ethical guidance on this matter, what about philosophy? An opposite view to arguing from fixed principles of right or wrong is to judge ethical issues with ideas put forward by Wittgenstein (1889–1951) in his *Tractatus logico-philosophicus* of 1921. He writes in section 6.4 that 'all propositions are of equal value' and that 'ethics and aesthetics are one and the same'. Saying that something 'is wrong' is the same sort of statement as saying that something 'looks ugly'. Whether this inner feeling is innate or learned from experience and education is not really the point. Individual psychology will help to determine whether people feel some action to be wrong or not, but there can be no valid objective test for its correctness or truth. Impersonal judgements in ethical matters therefore become impossible. One of the clearest British exponents of this position was the logical positivist A. J. Ayer (1910–89). To him all ethical judgements could be reduced to matters of emotional response. The world is devoid of fixed principles. To say that 'cloning humans is fundamentally wrong' is only an emotional expression that one has a strong feeling of disapproval about doing it, the so-called 'yuk-factor'. Another man may disagree about the wrongness of cloning in as much as it does not invoke the same feelings of dislike or disgust. When it comes to settling a dispute between the two there is no objectively valid way of determining whether this man or the other is either 'right' or 'wrong'. The argument should not be about questions of value, but should come down to a question of consequences. Will cloning lead to a clinically damaged child? Will the child be confused about his self-identity? Will the child be deprived of usual family relationships? Can the couple have a child by any other means? Does cloning fall outside the 'natural laws' of reproduction? And so on. Answering these factual questions may help to resolve the dispute.

Following these ideas on the relativity of ethics through to their logical conclusion seems to imply that anyone can justify anything if they feel good about it. Few, however, are willing to maintain this extreme view consistently. They would then have to deny that the actions of a Ghandi are no more different in kind than those of a Goebbels. They could condemn as immoral all the doctrines of such diverse teachers as Confucius, Buddha, the Hebrew prophets, Socrates, Mohammed and Karl Marx. Almost all human discourse would become meaningless if all were to hold that one ethical judgement is just as tenable as another. Few people would disagree with the value judgement that torturing innocent children is wrong in itself, regardless of any of the factual consequences. Ethical relativism does not appear to deal with fundamental issues like this with any conviction.

This, of course, constitutes no logical proof that there is an element of objective truth in ethical judgements. If the subjective view of ethics is upheld it can explain why there have been such very great differences that are found in adopted values. Ritual cannibalism and headhunting were sanctioned by New Guinea tribesmen but proscribed by Western Europeans.

Arguments from utilitarianism
Moral philosophers like to devise theories and arguments about how we citizens ought to deal with our difficult problems. But ethics should not simply comprise a catalogue of acts or a set of rules to be applied like a cookbook recipe. What use is a list of great ethical commandments if you cannot remember them just at

the time when you need them? Or what if you think of them in the wrong order? Why, like an alphabet, should they be taken in a particular order anyway? The greater need in ethics is for specific methods of enquiry to locate and overcome future difficulties such as those anticipated with the new eugenics.

What actually happens in practice? In my own line of medical work a more utilitarian approach is taken. There are a few primary guiding principles, the key ones being to prevent or cure disease and not to harm the patient. It is not possible to prove that health and life are desirable ends in themselves, they are assumed to be so *a priori*. But such principles are not rigidly fixed, mainly because of the complex nature of the medical situations that can arise. 'Thou shalt not kill', but the life-support machine is turned off under special circumstances such as severe and irreversible brain damage.

We undertake extensive observations of the sick person and then try to interpret the findings without bias or partiality with regard to race, colour or creed; and without bias for any one kind of diagnosis. The diagnosis has to be consistent with the external evidence. We have another guiding but flexible principle, that of intellectual honesty in the assessment of the facts. But we sometimes tell half-truths to patients if we think it might help their morale (it is called 'compassionate deceit'). Close relatives are consulted about this approach. However, a recent survey in Ireland found that about 80 per cent of cancer patients actually want to know the bare truth, whereas only 55 per cent of their relatives wanted the doctor to tell the patient the truth about their disease. So perhaps one should not attempt to predict the patients' preferences by talking to their relatives. If there is a chance of cure, but the treatment is extremely unpleasant, it is difficult to justify giving it unless the patient is told the absolute truth about their condition. So there is a variable scale of aims and values. These principles, values and criteria are not final ends in themselves but more like intellectual instruments for analysing the data of the current problem. We know that treating pleurisy and pneumonia with penicillin is more effective than a hot poultice, but a hot poultice is better than nothing. Some of these value judgements do become more important than others. To save a life in times of chaos or revolution, it may be not only permissible, but even a duty, to steal or take by force the necessary food or medicine for the patient. So far the National Health Service in the UK has not quite been reduced to this pass.

The future conflict of issues for eugenics, such as embryo enhancement or designer babies, could be resolved by taking the more pragmatic or utilitarian approach so often favoured by medical scientists. If the technique works and does not harm (admittedly a value-laden term) the child in any way, then the procedure could be allowed on the basis of expediency if any reasonable members of the public think it is in their interests to undertake it. Even if the vast majority of people feel that the technique is abhorrent according to their own moral code – either being unnatural or constituting a wrongful exercise of power by the parents over the welfare of their future child – the remainder should perhaps still be allowed to use the technology. After all, we allow a mother to continue with a pregnancy knowing that she is going to have a child with Down's syndrome. She may count it as a gift from God, even though we all know the child will be born deficient in some ways.

On this basis the ultimate sanction of the pragmatic approach (external motives apart) is whether it will produce the right outcome for the individuals concerned. This, of course, will have no binding efficacy on the rest of us who do not possess such desires and it may even provoke our hostility; witness the ongoing conflict over the simplest of eugenic techniques, namely abortion.

Arguments from opinion
Past decisions or old principles can never be wholly relied upon to justify a future course of action, particularly if more knowledge has been acquired. Medical ethics has to be protected from falling into the formalism of rigid repetition. It is rendered a flexible, vital and growing subject by continuous debate. Its values are created in the actual process of solving the immediate problem and not on

some set of historically formulated principles. To suppose that we can make a hierarchical table of values that will serve once and for all is a matter for armchair philosophy. Every clinical situation, such as a woman's request for termination of pregnancy, or posthumous insemination, requires individual analysis and an individual response. All our secondary standards, principles and rules lose their pretence of finality. They are just instruments to test and dissect the present situation. We temporarily follow some of the rules if we wish to achieve particular ends. (There are excellent reasons for following the rules of hygiene if we wish to preserve health, yet sometimes we will sacrifice these rules for the sake of temporary pleasures that in retrospect may be regretted and seen to be foolish.)

Public opinion
In democratic as opposed to autocratically ruled countries, public opinion may be the final arbiter for accepting the new eugenic techniques. The battle for public opinion will be waged on various fronts. The major battle lines will be drawn up by numerous adversaries: commercial companies using as their main weapon the dynamics of the marketplace; committees of experts in the fields of genetics, medicine, sociology, theology, etc., producing more or less factual and well-reasoned reports; the creative artists (of whom Shelley said that the poets should be the unacknowledged legislators for us all); and the media which are a variable mixture of all the others. Novels, films and the news media, amongst other sources, provide further opinions which influence the public.

Of the imaginative arts, three novels in particular, Huxley's *Brave New World* (1932), Orwell's *Nineteen Eighty-Four* (1949) and Mary Shelley's *Frankenstein* (1818) have been very influential in shaping public opinion. These books will continue to inform public opinion. All three of them have at their heart a distrust of scientists and fear of scientific progress. Another major theme is the distrust of the use of scientific technologies by unscrupulous politicians to establish a totalitarian state. In *Nineteen Eighty-Four*, Big Brother developed information technology whereby two-way

television screens instructed people what to do and also monitored their behaviour and thoughts to make sure they complied
with the orders without dissent. Aldous Huxley used reproductive
technologies to produce a tractable and uniform society. Another
theme is that of a scientist creating something that he cannot control, as brilliantly portrayed in *Frankenstein*. Man, having the
presumption to create new forms of life, was punished for his
hubris by the monster turning against his creator and killing all
those closest to him. The word 'Frankenstein' has even become
part of our vocabulary, defined in the *Oxford English Dictionary* as
'a thing that becomes formidable to the person who created it'.

Such imaginative literature has a far greater impact on public
opinion than the scientific literature or the pronounced arguments of the experts. The latter characteristically form
self-elected committees of reasonable experts including repre-
sentatives of all the interested fields – theologians, geneticists,
ethicists, physicians, psychologists and sociologists. They take
the problem in hand, give it a thorough scrub down, sterilise the
emotional content and compose a document dealing rationally
and logically with all the issues. But it does not really grip or
inspire the imagination of the public. The public is then told to
trust them and subsequent legislation may enforce the committee's recommendations. The public often believes that it has little
or no access to the committee or board meetings where the
decision-making occurs. We are beginning to mistrust the specialised guidance of the experts in their manifold schemes. We are
often unmoved by the flat ephemeral documents and the outcome of their learned deliberations, even if they are models of
logic, lucidity and sanity.

If one were to search for the source of public opinion in
Germany underlying the origin of the Holocaust one would end
up with a raft of superficial and half-baked accounts of large meaningless concepts from the so-called experts on the purity of racial
groups, mixed up with a mystical German philosophy on the role
of the *Ubermensch*. So much can bad opinions lead to disaster.
The poet John Milton wrote that the opinions of good men are

knowledge in the making. Even most of what we think of as knowledge may be no more than opinion. The opinions of a single highly gifted individual who can inspire the rest of us are of inestimable value. The leadership of an individual and his ability to influence public opinion outweighs all the work of committees whose efficacy has been said to be measured by the IQ of the highest participant divided by the number of people seated around the table. But which individuals should we turn to? Moral leaders of the past have little to say about the pressing genetic and eugenic problems of today. And anyway we further deify the global economy driven by the new information technology. Banks are far taller than cathedrals. Our prophets (and our profits) come from prices, wages, output, demand and jobs.

Personal opinion
My personal requirements for leadership in the field of eugenics would be an individual who has been trained as a scientist, can handle and interpret scientific data, and who publishes reputable scientific papers. Secondly, it would be helpful but not essential for the person to be medically trained to understand at first hand the health needs of individuals and their families. Thirdly, and most important of all, the person would need to have the ethical and aesthetic sensitivity of a great artist. Such people are exceedingly rare. The fact that there are few distinguished practitioners who would meet the above requirements suggests that, were he to be found, his opinions would be seen as a guiding light to us all.

One or two individuals from the past fit the bill. Anton Chekhov (1864–1904), the celebrated Russian playwright, also published scientific papers with tables of observations of demographic data and hypotheses about the socio-medical problems on the convict island of Sakhalin. He displayed the accuracy, precision and detachment of the field scientist. He also worked as a general physician until pulmonary tuberculosis forced him to give up his medical practice. His plays deal with contemporary eugenic ideas about the inheritance of personality traits. For example, in

Chekhov (left) discussing literary matters with Tolstoy.
(David King Collection)

Ivanov the main character who is hopelessly neurotic himself warns a young girl, Sasha, not to marry him saying: 'You will be ruining the human race. Thanks to people like you there will soon be none but neurotics and psychopaths born into the world.'

Of great interest is Chekhov's opinion, recorded in many letters,

on how far we should trust science and technology. He lived
through the second half of the nineteenth century and his period
in many ways reflects our own. He saw bacteriology do for infectious disease what genetics is now doing for inherited disease.
Great scientific advances were made during his lifetime. In 1862
Louis Pasteur had proved that germs such as bacteria can cause
disease and he originated and successfully used vaccines to protect
sheep from anthrax and (in 1885) human beings from rabies.
Robert Koch discovered the bacterial origins of tuberculosis in
1882 and cholera in 1883, leading eventually to a more rational
management for these diseases. In Glasgow Joseph Lister, following on from Pasteur's work, introduced the use of antisepsis in
surgery that, combined with the new developments in anaesthesia,
opened up the whole era of modern surgery. The ability to remove
body parts at will must have appeared as miraculous as cloning is
nowadays.

Chekhov followed these outstanding developments with the
greatest interest and commented on them in many of his letters. It
is worth considering some of those opinions expressed in his letters that have a bearing on our current problems with eugenics.

To start with one of his crucial points: 'that it is not my opinions
or convictions, however passionately held that are so important, as
the methods by which I arrived at them'. What would he make of
the fanaticism of the anti-abortionists, or the antagonists to the
genetic modification of crops? The methods by which individuals
arrive at their decisions are just as important as the final choices
made. How ambivalent are their motives? Are the choices founded
on the basis of pure self-interest or vanity; or is consideration
given to the best interests of all concerned? Are their choices
founded on religious or ideological dogma; or do they derive from
the life experiences of the person involved?

Chekhov goes on:

Science and technology are passing through a great
period now. The Russian people are living through an
infatuation with the biological sciences and a materialistic

movement will once again predominate. The biological
sciences are working wonders now, but they may advance
upon the public like Mamai [the supreme Tartar ruler of
the fourteenth century who laid waste a large section of
what was to become Russia]. However, acquaintance with
the biological sciences, combined with the scientific
method, always keeps me on my guard and I try when-
ever possible to bring my ideas into harmony with
scientific data.

In one final letter he replies to a difficult ultimatum between science and the humanities: 'My dear man, if I were offered the
choice between the scientific progress of the sixties, or the work
done in the most wretched District Hospital, I should without a
moment's hesitation choose the latter.'

Chekhov seems to believe, when there is a fundamental dispute
about the use of the new technologies (such as cloning nowadays),
that humanitarian values as exemplified in the work of a simple
country hospital should always prevail over and above the applications of the new scientific developments whose benefits are
largely untested.

This suggests that, given a choice, Chekhov would prefer to do
without the publication of Darwin's *The Origin of Species* than
not have the scores of Beethoven's nine symphonies. Many
people might agree with him. Darwin's theory of evolution
would probably have been proved anyway in the work of the next
generation of biologists; but would there ever have been another
Beethoven?

More about choices
The Swiss recently tried to impose a similar stark choice about the
use of genetic engineering in medicine and the food industry.
They used a public referendum to give the electorate an opportunity to express an opinion on a controversial topic, thus providing
a type of direct democratic choice. Referenda may be particularly
useful for 'hot potato' issues concerning moral or lifestyle choices

that cut across the usual party lines since all citizens are consulted directly on the topic at hand.

However, critics of the use of referenda suggest that citizens may well cast their votes in ignorance without understanding the real issues involved, or that a low turnout might make a victory seem illegitimate unless a reasonable number of eligible votes are cast to approve a proposal. Also, referenda will tend to ignore or brush aside the concerns of minority groups. Nevertheless the social applications of the new genetics would seem to be excellent topics to settle by the use of referenda.

The questions in the Swiss referendum were: 'Do you agree with the research and applications of genetics for daily life? Do you agree with their applications for clinical medicine? Would you agree to eat chocolate with genetically modified soya added?' Only about 40 per cent responded to these questions and the Swiss voted by a 50 per cent majority to retain the use of genetic engineering and the genetic modification of animals, plants and other organisms. Doctors had stood shoulder-to-shoulder with university medical faculties to lobby for the rejection of the ban on using the new genetics provided guidelines were put in place to protect the public interest. Lobby groups tried to persuade the public to follow their own principles and opinions in voting, and provided an educational forum to justify their views. Some of the large drug industries in Basel, such as Novartis and Hoffman La Roche, even made plans to remove their operations to neighbouring countries in the event that the ban on genetics was enforced; their research work would be unable to continue competitively without the use of the new genetic techniques.

In Switzerland and some American states referenda are a central feature of the political system, rivalling the legislature in significance. The Internet should make electronic referenda more widespread and thus involve the public much more in the decision-making process.

Final comments

The problems and issues that arise from the application of the new eugenics will make ethicists of us all. It is certainly pleasant and comfortable to have someone else work out the problems for us and then to keep to their prescribed rules of conduct, but the authority of experts has lost much of its appeal. Too many disasters have been committed in the past by the experts in the name of eugenics. When any government or pressure group tries to prevent people from doing one thing, or to force them do another for their own good, then one has to beware – the notion of liberty is being eroded. Such coercion has come in two main waves in Western Europe during the twentieth century: first in the 1920s and 1930s in the form of the old eugenics and social engineering; and then in the 1940s after the war, when economic planning and income redistribution attempted to enforce a form of social equality. Since then the waves have receded but the forces behind them are still very much alive.

It is for us to debate and feel our way carefully through all the eugenic advances that are to come. Wherever practically possible people could be left to make their own choices as to how the new technologies should be used. Coercion by experts, politicians or pressure groups often implies an unjustified claim to knowledge and certainty about its applications. The more that governmental or regulatory agencies interfere and meddle with such intimate and personal matters related to the new eugenics, the more likely are problems to arise by ill-thought-out and repressive regulation or legislation. With goodwill and good intentions and a combination of altruism and advice from experts and professionals, we should be able to arrive at the best applications and outcomes for society, to encompass the physical, emotional and mental well-being for the greatest number of people and equal opportunity and social justice for all.

There are no ready-made answers. We learn as we go along. The facts have to be respected but we must realise at the same time that our observations and theories are always incomplete. We need the intellectual humility to realise that just as certain colours of the spectrum, or several notes of the diapason, remain beyond our perceptions, so other means of existence may remain beyond our comprehension until we invent the instruments to explore them properly. We learn from our past mistakes and have to cultivate our own sense of right and wrong. Our current ideas and prejudices are constantly being tested and challenged by the new scientific theories and observations. And we have to learn to mete out our own form of justice as these new issues arise, to build a civilised, articulate and well-adapted community.

PART 2

Which Genetic Markers?

11

The Genetic Components of Disease

The spotted goats shall be your wages, and all the flock
bore spotted young; . . . The striped ones shall be your wages,
and all the flock bore striped young.

—Genesis 30:31

Single gene disorders

In a new foreword written in 1946 for *Brave New World* Aldous Huxley stated that a totalitarian government will require, amongst other things, 'a foolproof system of eugenics designed to stand-ardise the human product and so facilitate the tasks of the managers'. One aspect of standardising the 'human product' is to eliminate genetic variation causing disease. Most, but not all, people would consider this a worthwhile aim.

There is a wide spectrum of disease. At one end, genes have almost no effect on the appearance of the disorder, whereas at the other end the genes are the major causes. A healthy man or woman crossing a street in New York gets hit in the head by a stray bullet from a policeman who is chasing someone else. The resulting injury is determined purely by a chance environmental circum-stance and is not genetic at all.

Now consider a newborn child who looks normal but who fails to pass the developmental milestones of crawling, walking and speaking properly. If left untreated the child becomes mentally defective. This can be due to a purely genetic disorder. In this example the defect could be a change in a single gene that makes an enzyme for the cell that is involved in the breakdown of one of the components (the amino acids) of the protein foodstuffs in the child's diet. The child is severely maladapted to eating protein food and as a result an abnormal chemical builds up in the bloodstream to interfere with the brain's development. One such disease is called PKU (standing for *PhenylKetonUria* – the abnormal chemical in the urine). It is a very rare mutation affecting only about one in 10–15,000 new births in Europe or America, but it is so serious if left untreated that all newborn babies are tested for its presence. If they are found to have the mutation they can be treated by simply restricting the dietary components that they cannot handle properly. PKU was one of the first genetic diseases to be screened in early infancy by taking a pinprick of blood from the heel of the newborn. Dietary treatment in early infancy is very effective. If the mother is known to be a carrier of the abnormal gene she will also have to be on the diet while she is pregnant otherwise the raised levels of the abnormal chemical in her blood can induce mental deficiency in her developing foetus.

There are many other diseases caused by single gene defects and some of them are listed in the Appendix, Table 3. Fortunately, they are all rare and screening tests have been devised for some of them so that they can be detected early and the proper treatment started promptly.

Haemophilia is a genetic disease of the clotting mechanism of the blood where one of the essential proteins is not made properly. The mutated gene is located on the female sex chromosome (the X-chromosome). Because males have only one of these they cannot make the clotting protein if they inherit the defective gene from their mother, but the female, who has two X-chromosomes, can make enough of the protein to stop her bleeding with just one good X-chromosome – hence she is called a carrier.

In the words of one twentieth-century victim, the disease is an 'everlasting bloody nuisance'. The problem is that the sufferers cannot stop bleeding after the slightest injury. The smallest movement or damage can cause bleeding from the nose, mouth, gums, skin and most painfully into the joints. Queen Victoria was a carrier of the disease although it is not known how she acquired it. From genealogical studies she did not appear to inherit it, so it was either due to a new change (mutation) in the letter sequence of her gene coding for the particular clotting protein (very unlikely, with a probability of its occurrence as a chance event being about one in 50,000) or to the possibility was that she was the illegitimate child of a man carrying the bad gene (but no one has estimated the chances for this). Two of Victoria's daughters (Alice and Beatrice) possessed one copy of the bad gene that in itself does not cause any harmful effects.

Although haemophilia cannot be treated as casily as phenylketonuria one can replace the missing protein (and perhaps soon in the future the defective gene; see page 44) for the blood-clotting mechanism and help to staunch any source of blood loss. Haemophiliacs have to be very careful to avoid any situations that would give rise to bleeding. These would include dental treatment, minor surgery and sports or outdoor activities.

Another single gene disorder, already discussed from a social viewpoint in Chapter 8, is sickle cell anaemia. The defect here is in one of the genes coding for the pigment of the red blood cells, the haemoglobin. As a result the red cells are unstable, particularly when the oxygen content of air is low. The haemoglobin forms into long strands and deforms the shape of the red cells into sickles rather than being perfectly round as they are when the haemoglobin is normal. These sickle cells aggregate together and can block small blood vessels and lead to blood clots anywhere in the body. Sufferers have to avoid situations where the oxygen levels are low, such as living at high altitudes, mountaineering or flying in planes, because of poor air quality.

Notice that even with these single gene disorders there is often an environmental factor that needs to be present before the disease becomes apparent – examples include dietary protein, minor injuries or low oxygen content of the air. Knowing the environmental factors that interact with the genetic disorder

offers in some cases a way of modifying the progress of an inherited disease. So there is no need to be fatalistic about all inherited
disease; its severity can be greatly ameliorated by attention to the
environment.

Common genetic diseases

There is a group of common diseases occurring in more than 1 per
cent of the European and American populations that clearly has a
genetic component but this is not sufficient by itself to cause the
disease. These include cancers, diabetes, heart attacks, strokes and
the premature dementias. The genes only predispose people to
develop the disease if they are challenged by a particular environmental factor. For example, there are several genes that predispose
a person to develop a heart attack but this occurs only if too much
fat or cholesterol is consumed in the diet, or if the person smokes
excessively, or has additional problems such as high blood pressure.

If the environment plays such a large part in the cause of the
disease, how can one be sure that genes are actually involved too?
The best evidence comes from family studies where the disease
can appear in many members of the same family – but one could
argue that perhaps they shared the same environmental factors
since they were brought up within the same family background.
Better evidence comes from studies of twins.

There are two types of twin: 'identical' ones (technically called
mono-zygotic twins), who have exactly the same genetic constitution as each other; and 'non-identical' (called di-zygotic twins)
who share the same number of genes as any pairs of brothers or
sisters. If one studies pairs of identical twins and one of the twins
develops a disease such as diabetes, what are the chances of the
other twin developing diabetes? If the disease is purely determined by the genes, all the other unaffected twins at the start of
the study should go on to develop the disease. This is called a
concordance rate of 100 per cent, meaning that both twins get the
disease or other characteristic that one is studying. In fact for one

type of diabetes the concordance rate is about 60 per cent, suggesting that not all of the disease is caused by the genes. Of course, if the twins are brought up under the same roof they might have encountered the same environmental factor that caused the disease. One can then study a comparison (or control) group of pairs of non-identical twins who also share the same family background. If the concordance rate for diabetes of the non-identical twins is only 30 per cent, it means that the environment is not playing that large a part. It would be a 100 per cent if the environment were the sole cause.

Even stronger evidence can come from identical twins who are brought up separately, perhaps because they have been adopted by different families or been placed in different orphanages. In this situation the environments of the twins can be very different, but if both twins develop the disease in more than 60 per cent of the pairs, this would be very strong evidence for the inherited nature of the condition. One can arrive at a measurement, called 'heritability', which attempts to show how much the genes contribute to the disease and how much is contributed by the environment. The measurement of heritability is calculated from the concordance rates of the disease in identical and non-identical pairs of twins; and the values for some common diseases are presented in the Appendix, Table 4. Unlike the rare single gene disorders described at the beginning of the chapter, these common diseases generally have many different genes that can underlie their inherited basis and this makes screening for them much more difficult.

Screening for genetic disease

We have already seen how screening for the genetic disease phenylketonuria (PKU) is undertaken at birth and how the introduction of the correct diet to a baby of less than four weeks old can prevent intellectual impairment. This was one of the first clear demonstrations of the value of genetic screening, provided there are satisfactory treatments available for the affected children.

Nowadays, however, we do not have to wait for the birth of the child to screen for genetic disease. We can test embryos at an early stage in their development in the womb to see if they carry a genetic disorder and then, if there is a serious problem, offer the mother a termination of pregnancy after explaining to her the full details of the baby's condition and obtaining her full consent. To make the diagnosis involves an uncomfortable procedure using syringes and needles to take a small sample of the placenta for analysis. Soon it will be possible to obtain the same information from a sample of the mother's blood in which a few individual foetal cells can be found from which to extract the DNA.

A good example of screening for genetic disease is that undertaken for Down's syndrome. Strictly speaking this is not a single gene disorder but more a disease resulting from the chromosomes failing to separate properly during nuclear division. These are the structures found in the cell nucleus that transport the genes to each of the daughter cells at the time of cell division. They are a sort of packaged transport system. In Down's syndrome, in addition to the usual paternal and maternal chromosomes being transferred into the fertilised egg at the start of the development of the embryo, an extra chromosome is accidentally transferred as well. The resulting baby will possess three instead of two copies of a chromosome (usually chromosome number 21) in most cells of its body. This impairs the child's development in many ways, including defects such as stunted growth, mental impairment and blunted facial features.. The presence of the three chromosomes can be detected at an early stage in pregnancy by collecting some of the fluid surrounding the developing child in the womb and examining the cells taken from the baby by using special microscopic techniques. If the baby carries three copies of chromosome 21 the mother, after appropriate counselling, can be offered a termination of her pregnancy.

These are the principles for detection of all the single gene disorders listed in Appendix Table 3, except that instead of examining for chromosomal abnormalities one can directly test for the presence of mutated genes that transmit the disease by the use of genetic markers, as described in Chapter 2. Methods are now being developed to detect thousands of genetic markers at a time in DNA samples taken from a single cell of an embryo after *in vitro* fertilisation. This is before implantation into the mother's womb. The embryo can then be discarded if it contains a 'bad' genetic make-up and another embryo created by test-tube fertilisation can be used in its place.

Brave New World was published in 1932 but Huxley set it 600 years in the future because he thought the ideas in it had 'been pushed to fantastic, though perhaps not impossible, extremes'. In fact, sixty-nine years after the novel was first published we can now do almost all the things he describes except hatching the fertilised egg all the way to a full-term baby in artificial containers. The speed of acquiring this new knowledge would have astonished Huxley. The egg can be fertilised by sperm in glass dishes and the embryo developed up to the stage of a ball of cells *in vitro*. Embryos at this stage can be screened for a variety of common conditions using genetic markers. They can be classified according to Huxley into Alphas, Betas, through to Epsilon embryos depending on their future level of capabilities. We can make multiple clones of, say, the Epsilon foetuses, following the techniques of Dolly, the cloned lamb, to make armies of identical twins to carry out all the menial tasks in society. We can make the 'eyeless monsters' Huxley writes about, but for which he could find no use. Animals without heads have been reared (anencephalics), which, if human, could be used in Huxley's hatchery on the conveyor-belt system for incubating embryos up to the time of birth.

Needless to say the society that Huxley imagined was horrific, being a satire on totalitarian regimes and a warning against the dehumanising aspects of scientific progress. Scientific applications are always double-edged swords; they can be used for either benevolent or malevolent ends. The next chapter considers some of the more beneficial uses of genetic markers.

12

Frail New World

When a lot of remedies are suggested for a disease,
that means it can't be cured.

—Anton Chekhov, *The Cherry Orchard*

There are no short cuts to knowledge despite computers telling us that all we have to do is simply point and click on our targets. Scientific research often involves going up alleys only to find out that they are blind. One constantly has to risk absurdity; and I personally prefer the absurdity of doing medical research than the even greater absurdity of not doing it. Scientific method is still one of the soundest ways of acquiring new knowledge.

Sometimes medical phenomena seem to be irreducible to any system, pattern or explanation. The raw data resemble a kaleidoscope inside a magic lantern, but painstaking experiments are slowly replacing vague speculation. Laborious observation, interpretation and reinterpretation of data, then repetition and confirmation of experiments are all needed to ensure accuracy. All these have led to a sound and detailed contribution to the knowledge of some of the harmful and beneficial genetic variants that occur in embryos, children and adults.

Before genetic markers were discovered we had to search for abnormal constituents of the blood or urine that result from the effects of the defective gene. Now we can take cells from the blood of an adult or child, or from the fluid in the womb that surrounds the embryo, or even a small piece of the placenta that is derived from the embryo, extract the DNA and analyse it directly for the genetic variants. This is already being done for the rare group of genetic disorders listed in Appendix Table 3. Screening for such disorders can have a substantial impact on the families that harbour such bad genes, but since the conditions are uncommon (they occur in about 1 in 10–50,000 births) it has little impact on society as a whole. However, screening for a predisposition to the common disorders of cancer, heart attacks and early dementias could have many consequences on health care planning, the wellbeing of our newborn children and other social matters such as life insurance and employment.

Genetic screening can create two new classes of people: the 'asymptomatically ill' – people who are at risk of developing a disease in the future but who are quite well at present – and the 'worried well', those who do not carry the disease marker but who are not reassured by the tests. The public response to the use of genetic screening can vary between two extremes: 'People need to know about their genes and be aware of potential diseases that may occur' or 'it's all a bit frightening. It makes you wonder if you want all this knowledge anyway.' What are the possible diseases that may occur, and how far does the knowledge of, and research into, them extend? To illustrate the principles of the research going on in this area, it is worth considering in detail the three common diseases: heart attacks, cancers and dementias.

Can you inherit a heart attack?

Heart attacks are the commonest cause of death in middle-aged men in northern Europe and the USA. They account for more deaths than all the cancers added together. A heart attack occurs when an accumulation of cholesterol-rich fatty material, called atheroma, in the vessel walls of the arteries of the heart obstructs blood flow. The heart muscle beyond the obstruction becomes starved of oxygen and this produces a cramp-like pain across the chest, called angina. If the shortage of oxygen is prolonged the muscles of the heart die. This can interfere with the efficiency of the heart's action as a pump and can produce heart failure. Sometimes the heart can stop beating altogether, causing sudden death. The range and severity of the symptoms depend on how many of the blood vessels of the heart are affected by atheroma and how extensively and for what length of time the heart muscles are deprived of oxygen. Three major predisposing genetic – and therefore heritable – disorders that help to cause heart attacks and strokes are raised blood fats; raised blood sugars as in diabetes mellitus; and raised blood pressure.

Raised blood fats
Convincing evidence that heart disease can be caused by genetic factors comes from studies of a rare single gene disorder, familial hypercholesterolaemia. This occurs in the UK and USA in about one in every 500 births. In this disease one of the fats circulating in the bloodstream, cholesterol, cannot be transferred into cells properly because of a defect in a channel guarded by a gate on the surface of the cell (technically called a receptor). This gateway is responsible for the removal of cholesterol from the blood into cells; if it is defective the levels of blood cholesterol will rise. Cholesterol can lead to deposits in a variety of sites including the skin around the eyes, in muscle tendons and most dangerously in the arterial walls of the heart, causing the accumulation of fatty atheromatous plaques. This can then lead to a heart attack with complete obstruction of the blood vessel. A clot forms on the surface of a damaged fatty plaque to block the artery. A person unfortunate enough to inherit two doses of the bad gene, one from the mother and the other from the father, is liable to develop heart attacks from a very early age. The earliest recorded case of a heart attack is in an eighteen-month-old baby. The child was found to have two copies of the defective gene, one from each parent. This is proof positive that a gene defect can cause a complex condition such as a heart attack, but it is extremely rare.

Are there other genetic factors that could predispose to the commoner forms of early heart attacks? Yes – the evidence for this comes from studies of identical and non-identical twins, as explained on page 204. The occurrence of similar features of an early heart attack before the age of fifty-five is found in forty-four out of 100 identical twin pairs, but in only fourteen out of 100 non-identical twin pairs. From these figures one can calculate a value called the 'heritability' that comes out at 0.65. This means that deleterious genes could account for about 65 per cent of the cause of the disease, whereas the other 35 per cent would be due to environmental factors such as smoking, lack of exercise or poor diet. The heritability value for heart attacks diminishes as the age of onset of the disease increases between the seventh and the eighth decades of an individual's life, suggesting that environmental factors become more important as the person grows older.

There are several reasons why it would be useful to know what these genetic factors are. They do not actually cause the disease but only predispose a person to develop the disease if he or she encounters the relevant environmental conditions. Thus, if one knows the genetic risk factors in a person one can then modify or alter the relevant environmental factors. The term 'risk factor' in this case means some measurement that will predict the outcome of interest, that is whether a person will develop a particular disease.

Consider a simple case: if one of the genetic variants were to be found at the gateway that removes cholesterol from the bloodstream, reducing the dietary intake of cholesterol would reduce the load on the defective cell gate and correspondingly lower the levels of circulating cholesterol in the bloodstream. This in turn may lead to reduced deposits of cholesterol in the arterial walls of the heart. Many patients with heart attacks have such defects and are asked to go on to low-cholesterol diets. However, people are often unwilling to change their lifestyle, either by going on a diet or stopping smoking. Doing the latter would improve health. Quite apart from cancer and heart disease caused by smoking, it is likely that it damages the lining of blood vessels and makes them more permeable to the entry of cholesterol.

If it were as simple as just one gene being involved in predisposing to premature heart attacks before the age of fifty-five, the disease might have been eliminated many years ago by preventive measures, as in the case of phenylketonuria (PKU). Unfortunately there are many more genes than just the one involved for the development of the fatty plaques in arteries: at the last count there were more than forty contributory genes. These include genes making proteins involved in fat transport in the blood; genes that predispose to a high blood pressure; gene defects that cause a rise in blood sugars leading to diabetes; and genes that make the blood more liable to clot on the surface of the fatty plaque of the arterial wall. And each gene may have only one out of hundreds of different possible mutations that predispose to the disease. The problem becomes much more complex than looking for a single letter change in the code of a single gene.

Raised blood sugar: diabetes mellitus
This is a common and very old disease, descriptions of which were first found in Ancient Egyptian papyri. It used to be called the 'pissing evil': victims pass large quantities of urine and gradually waste away by converting their body proteins into sugar which they lose in the urine. There are at least two types of sugar diabetes. The first is insulin-dependent diabetes which must be treated with the hormone insulin and was fatal before the discovery of insulin in 1921. The second, non-insulin-dependent diabetes, is due in part to the inefficient action of insulin in the body, particularly in muscle, fat and the liver. In terms of health economics non-insulin-dependent diabetes costs Britain's National Health Service about £2 billion a year, about 4 per cent of the total annual NHS expenditure. Early diagnosis and treatment of the diabetes would be expected to reduce the complications of the disease and hence the expenditure. This is where genetic screening would come in to help to identify susceptible individuals.

Both types of diabetes are caused by a combination of genetic variants that interact with various environmental factors such as viral infections, damage to the pancreas, stress and obesity. The disease runs in families and pairs of identical twins are much more likely to have it than non-identical twins if one of them is initially affected, indicating that genes are involved. Different combinations of genes appear to predispose to the two different types of diabetes. In very rare instances the disease can be due to a single gene defect but this accounts for less than 1 per cent of all cases. More usually there are multiple genetic variants affecting how the body handles glucose or how cells making insulin in the pancreas are protected from damage. In insulin-dependent diabetes there arc major genetic variants on chromosome 6 that may be responsible for about 42 per cent of the inheritance of the condition. By contrast, in non-insulin-dependent diabetes no single major gene has been discovered and the disease is more likely to be due to a combination of variants, rather like the situation for heart attacks and strokes.

Our current knowledge of the genetics of diabetes is of little use in clinical practice because it is still incomplete. This is being improved every year. When all the susceptibility and protective genes are known it will be possible to map them for each individual. Rather than doing mass screening of the population for these genetic variants it may be more efficient to advise susceptible individuals, as detected from their family histories, to take the tests if they so wish. If the tests show positive for a majority of variants then preventative lifestyle measures (regulation of diet, exercise, alcohol intake, smoking) can be introduced and regular monitoring performed to detect occurrence of the disease at an early stage.

Diabetes causes a 30 per cent reduction in life expectancy. It is the commonest cause of blindness in the UK in people between the ages of twenty and sixty-five; and it is the commonest cause of kidney failure in developed countries, with its incidence varying by race and region. From the public health standpoint the only cost effective way of dealing with diabetes is to prevent it. In many cases it is already a preventable disorder. But when all the genetic factors are worked out it will become even more so.

Raised blood pressure

The blood is pumped round the body by the heart under a definite head of pressure. This can be measured by an inflatable air cuff around the arm attached to a mercury level that can record pressures. Like the pressure in car tyres, the pressure in blood vessels must be kept within strict limits, otherwise it can damage the walls of the vessels. The blood vessels may rupture and cause a haemorrhage, which in the brain can lead to a stroke, or in the back of the eye can lead to blindness. Alternatively, the blood vessels can react to the raised pressure by the wall becoming thicker and inflamed which can lead on to an accelerated form of hardening of the arteries.

In rare cases high blood pressure can be caused by a single gene defect involved in the regulation of the amount of salt and water in the body, which in turn affects the level of the blood pressure. More commonly the inherited component of the disease is contributed by the action of several defective genes that predispose to the development of the high blood pressure. A search is under way to identify such genes and some variants have already been implicated. The search is still at an early stage. It involves different types of genetic analysis such as population studies for association with genetic variants; the study of the pattern of inheritance of gene defects in families when two or more siblings have the condition; and a complete scan of all the genes in the body with the use of genetic markers to see if any of them links to high blood pressure. If found, they could be used to identify people who are at risk of developing high blood pressure in the future and who could therefore be monitored more carefully for the appearance of the raised pressure. The markers could also be used to predict which affected individuals are most likely to respond effectively to the different types of drugs that are currently in use. The implicated genes may lead to the discovery of new and more potent lines of drug therapy for treatment of the condition.

Can you inherit cancer?

Cancer is the one word patients dread when they undergo a seemingly endless set of tests in hospital. Of the cancers, after lung cancer those of the lower bowel are the commonest. In Britain about one in twenty-five people risks getting bowel cancer in their lifetime. It kills about 48,000 Americans annually. Any cancer is extremely unpleasant, and this one is no exception. It can obstruct the bowels, spread to the liver, bone, brain and other parts; only about 40 per cent of people will survive five years after the diagnosis has been made.

Bowel cancer

There is a particular group of people at high risk of developing bowel cancer: those who have two or more close relatives who have already had it, or one relative diagnosed with it before the age of fifty-five. The family history of the disease suggests a genetic predisposition, and some genes have now been incriminated for starting the cancer. One theory suggests that the victim inherits one bad gene from one parent; and then a second mutation occurs spontaneously in the patient to trigger the growth of the tumour. The faulty genes that have so far been identified are those involved in the repair of damaged DNA, and those involved in suppressing growth of the tumour.

Techniques to detect these genetic mutations are already being used for individuals in families where other relatives have previously been diagnosed with the cancer. If any are found to possess these harmful mutations they are regularly checked for the appearance of bowel cancer by the employment of a type of telescopic examination of the bowels (a colonoscopy). If an early cancer is found it can be excised immediately to prevent its spread. Using this screening procedure it may be possible to improve five-year survival rates to more than 70 per cent. If family members do not possess the harmful mutations they can be reassured and will not require such intensive check-ups.

When more of the faulty genes have been identified, and there are probably many such mutations, it may be possible to undertake population screening to narrow down the numbers of people at risk to a small group who will require annual examinations for cancer. It has been calculated that about 1,500 cancer deaths per year may be prevented by such measures.

Knowledge of the faulty genes will also help the development of new drugs to target the defective gene by either replacing the abnormal gene product, that is the protein, or even by reversing the genetic defect. A similar picture is emerging for other cancers, including breast and ovarian malignancies. A small combination of mutant genes appears to confer susceptibility to the cancer and screening the population for these may identify a small 'at risk' group who will benefit from intensive medical surveillance for prompt surgical action if the cancer develops.

Breast cancer

This dreadful disease kills about 44,000 American women and about 12,000 women in the UK each year. In both the USA and the UK if a woman lives to around ninety she will have a one in eight chance of developing this cancer (the risks increase with age). Any predictive test that could identify a woman at risk might save many thousands of lives by frequent screening for the first signs of the cancer and then doing a radical excision of the tumour at an early stage. One example of how this can work is shown in the medical histories of two sisters, Jackie and Emma. They recently took part in a research programme looking for mutations in one of the genes that confer susceptibility to breast cancer. They learned that they both carried the mutated form of the gene that put them at increased risk of developing the cancer. Subsequently a mammogram revealed that Emma had a small cancerous lesion in her breast but no cancer was detected in Jackie. Both sisters decided to undergo surgical removal of both their breasts. Jackie and Emma felt strongly that they benefited from knowing this genetic information. Many women have taken such desperate measures as soon as they are notified as being at high risk, so fearful are they of developing the disease. They have invariably seen a relative suffer either pain, indignity or even death from the disease.

There are several genes that offer some hope of a predictive test, called BRCA1 and BRCA2. The normal forms of both genes help to suppress the wild overgrowth of cancer cells that eventually spread to other parts of the body such as the lungs, the brain or the bones. There are more than two hundred mutations of these genes all having different effects depending on their exact position within the gene. Some of the effects can be to block the gene's normal function and this can lead to breast cancer. Since every woman has two copies of each gene, one from each parent, they are thought to inherit a bad copy from one parent (hence the family history of the disease) and then go on to develop a spontaneous mutation in the other gene that unleashes the cancer. This is a 'two hit' hypothesis, meaning that genetic changes need to occur in both parental copies of the gene before the disease will occur. It is the most likely event, as it would be very unlikely for a woman to develop two spontaneous mutations in the BRCA genes or to inherit two bad copies of the genes from her parents. Some mutants of these genes give a woman an 85 per cent chance of breast cancer by the age of seventy, compared with about 8 per cent in the rest of the population. Women with some BRCA2 mutations were found to develop fatal cancers as early as their twenties.

These statistics have come from family studies where there was a particularly high risk of developing breast cancer because at least four close blood relatives had this or related forms of tumour before entering the study. Such families have helped to identify the culpable genes in the first place. Other studies on larger groups have found weaker links with these mutant genes with risks more like 56 per cent by the age of seventy.

Notwithstanding these weaker statistics some biotechnology companies have already started to market tests for breast cancer genes. Myriad Genetics, based in Salt Lake City, Utah, will check a blood sample for all the known variants of BRCA. The company does, however, recommend that its customers first see a genetic counsellor who can explain the limitations and implications of the test (since few doctors have the necessary time or knowledge). Other companies or institutions are equally proactive. Aware that Ashkenazi Jewish women are more susceptible to a rare type of breast cancer than others, the Genetics and IVF Institute in Fairfax, Virginia, has directly targeted the Ashkenazim community even though such tests can make no definite predictions for any particular individual. The estimates of risk for cancer are calculated from epidemiological studies. They do not provide the complete information necessary to treat individuals. To rely on such tests to have one's breasts removed (an irreversible decision after all) is perhaps foolhardy but understandable. Clearly the commercial promotion and sales of such tests cannot be left to the biotechnology companies themselves and some form of regulation needs to be put in place to protect the public, since the issues can be so complicated.

Can you inherit dementia?

Bodily decay is gloomy enough, but even more abhorrent is a body without a mind. Among the catastrophes of ageing are the various dementias, of which Alzheimer's is one of the commonest.

The victims gradually and relentlessly lose all their memories which trickle away like water until nothing of the world remains for them. What a person is for the most part depends on memories. When words and meanings disappear, thinking is no longer possible. When the names of people, places and things, when the ability to recognise wife, husband, children, grandchildren or friends are all wiped out, there is nothing much left but the physical shell of the person.

In the early twentieth century an obscure German physician, Alois Alzheimer, was recruited by the great German psychiatrist Kraepelin, who was working in Munich at the time, to undertake a microscopic examination of the brains of people who had died with dementia. In 1904 Alzheimer was the first to describe the brain damage of patients dying of the condition then called 'general paralysis of the insane'. Later this was found to be due to infection with the microbe causing syphilis. However, not all cases of brain degeneration were so easy for him to classify. In 1905 Alzheimer described a new form of brain disintegration in a shrunken brain of a fifty-one-year-old woman, who had a progressive and severe memory loss. The changes were quite unlike those seen in syphilis and consisted of curious plaques and nerve tangles that we now know to be the hallmarks of a hitherto unknown dementia characterised by early memory loss. Kraepelin agreed to name it Alzheimer's dementia in honour of the discoverer.

Alzheimer's disease is a highly distressing and common degenerative disorder of the brain. It can attack up to 5 per cent of people older than sixty-five and about 30 per cent of those over eighty. There are more than 300,000 people affected by Alzheimer's disease in Britain and in the United States more than four million, resulting in over 100,000 deaths per year. If Alzheimer's disease occurs early in life a single abnormal gene is usually the cause, whereas in later life there may be several different genes that predispose to the disease. It is estimated that health care costs for this disease in the USA alone total more than $60 billion annually.

In Alzheimer's the damage consists of a tangle of filaments developing inside clumps of degenerating nerve cells and plaques of abnormal proteins accumulating in the grey matter of the brain. Over time, from months to several years, the grey and white matter of the brain gradually erodes. This can be visualised by the use of special imaging techniques such as CAT scans or MRI scans.

Although no definite environmental factors have been identified as contributing to the disease, several genetic components have been discovered. The evidence that genetics can play a major role comes from rare genetic defects of a particular protein (the amyloid precursor protein) that actually causes the disease at an early age, usually before sixty-five. More often the genetic effect is due to a combination of several mutations occurring in different genes that predispose to the degeneration of the brain. Surprisingly, one of them is involved in fat transport in the bloodstream.

This gene is called apo E and it makes a protein that acts as a vehicle to move fat around the blood. The apo E gene exists in three common variants called apo E2, apo E3 and apo E4. Everyone has two copies of the gene but the variants for each person can be different depending on inheritance from their parents. That means individuals can be E2/E3, E3/E4, E3/E3, or any other combinations. In Europe apo E3 is the commonest version of the gene, and more than 80 per cent of people have at least one copy of apo E3 and 40 per cent have two copies. However, people who have two copies of the apo E4 gene, that is about 7 per cent of Europeans, have a markedly increased risk of developing Alzheimer's dementia. More than 90 per cent of such people will develop the disease by the time they are seventy. About the only thing at present that can prevent this is premature death from another cause, which of course would usually have happened before the twentieth century when life expectancy was on average about forty to fifty years.

Even if you only have one copy of the apo E4 gene the chance of getting Alzheimer's dementia before the age of seventy-five is still high, about 47 per cent compared to about 20 per cent for people who have no copies of the apo E4 gene. The causal connection between the apo E4 gene and the development of the dementia is still a mystery but appears to involve an aggregation of the apo E4 protein with a group of other proteins (with such names as beta-amyloid, tau, ubiquitin and presenilins 1 and 2) to form the plaque.

There are many practical consequences of this disease. The problems of purchasing critical illness cover from insurance companies if you are at high risk of getting Alzheimer's dementia is discussed at length in Chapter 8. If you indulge in activities that raise the risk of an early dementia, then screening for any apo E4 genes that you may possess makes good sense. Repetitive head injuries resulting from boxing cause brain damage. About one in seven boxers gets a Parkinson-like disease or a premature dementia.

This is likely to be accelerated if you carry two apo E4 genes. It would be very interesting to speculate, for example, on the apo E4 status of a professional boxer such as Muhammad Ali. If he possessed two apo E4 genes he would have been well advised to avoid his 'Rumble in the Jungle' with George Foreman. He should have quit his profession early, while the going was good.

Footballers head footballs travelling at high speeds about seven hundred times each per year. This has been shown to result in considerable wear and tear on their brains, affecting memory and other cognitive functions. Again, if any footballers possess the apo E4 gene they should perhaps be warned at the start of their careers about the possible risks involved.

If you want to get the gene test done on your own you cannot order it. The company that markets the test, Athena Neurosciences, of San Francisco, will not examine your DNA unless your doctor thinks you have early signs of the disease or there are other pressing needs. This is because doctors consider that it is only an aid to making the diagnosis in people with suspicious symptoms. They believe that it should not yet be used for prediction of the disease in otherwise healthy individuals because there are too many other factors to take into account.

Following the discovery of such genes, new treatments are developing apace. The most successful therapy may be based on the knowledge of which genes are involved, by targeting drugs to correct or reverse the genetic defects in a particular individual. The large drug industries such as the recently merged companies of GlaxoWellcome and SmithKline Beecham believe in a fundamental shift in research from finding new drugs by chance to developing them systematically through an understanding of the genes that underlie the condition. The products of such genes may be modified by the development of new drugs for the benefit of the patient. For example, one new drug that treats the symptoms of Alzheimer's disease, Donepezil, is known to slow down the memory loss in dementia patients more effectively in those possessing an apo E3 or apo E2 gene than those with an apo E4 gene. Within the next few decades it is hoped it will be possible to screen for the many genes that influence the disease so as to make an early diagnosis before extensive memory loss occurs. Then it is to be hoped that preventative treatment, either by drugs or gene therapy, could be instituted to delay progression of the disease or even prevent it altogether.

Parkinson's disease

This is the second most common condition after Alzheimer's disease where one finds degeneration of the brain. It occurs in about 2 per cent of people older than sixty-five. The degeneration occurs in the part of the brain that controls bodily movements, and memory is not usually impaired at the start of the illness. The disease is characterised by rigid muscles, a tremor of the hands when they are at rest and difficulty in maintaining balance when walking. Sufferers have a characteristic shuffling gait and can easily lose balance resulting in many falls. Scientists at the National Institute of Health, Bethesda, USA, published a study of a large Italian-American family in 1996 where Parkinson's disease was diagnosed in relatives across five generations. They showed a gene on chromosome 4 was implicated that makes a protein called alpha-synuclein, and when the mutation occurs in affected family members the abnormal protein so produced aggregates with other proteins in the nerve cells to produce abnormal clumps called Lewy bodies. These observations have been confirmed in studies of several smaller families.

Although these genetic advances are not likely to result in immediate benefits for sufferers of this distressing disease, they may lead to the discovery of better drugs to control it once we know the genes and proteins that are involved in the damage to the nerve cells.

Screening: compulsory or voluntary?

Voluntary screening has already been adopted without government sponsorship by some groups of people. In the USA some of the Orthodox Jewish communities, such as the Lubavitcher community in Brooklyn, have established a programme to screen all young Ashkenazi Jews for the Tay-Sachs gene that produces a fatal degeneration of the nervous system. A database has been established for the results so prospective marriage partners can find out if they are at risk of transmitting the disease to their unborn children. Some children are tested anonymously and the results recorded on a confidential database. If two people of marriageable age are both carriers they would not be introduced to each other as prospective partners. Since marriages are usually arranged in this community this would not necessarily be considered peculiar. It is a form of community-sponsored eugenics but it could lead on to wider fields of control. However humane the intention of this programme may be it must be considered as a socially controlled rather than a voluntary eugenic plan. One organisation that promotes this means of combating Tay-Sachs disease is called 'The Association for an Upright Generation' – a clearly eugenics-inspired title. Is this plan a cruel and absolutely unwarranted intrusion into the private liberty and rights of the individual? I do not think so. It is a perfectly logical extension of the veto on closely consanguineous marriages that is at least partly based on genetic considerations. We all accept the veto enforced by law to prohibit brother/sister or father/daughter marriages. The Tay-Sachs example shows how the social perceptions of eugenics can change with the acquisition of new knowledge. With the future availability of a multitude of genetic tests, it is likely that potential marriage partners may themselves request screening to see if they both share mutually deleterious genes.

It requires sound judgement to really make this knowledge work properly in society. It raises the fundamental issue of how far we should proceed with eugenics controlled by outside agencies.

Such programmes may also lead to stigmatising minority groups. As discussed earlier (see page 218) there are genetic determinants that are found more commonly in Ashkenzi Jews for rare types of breast and colon cancer than in other groups. When Jewish bloodlines can carry diseases such as breast cancer, colon cancer or Tay-Sachs disease, there is a real risk of racial stigmatisation. Other groups may have just as many genetic links to a host of other diseases; it is just that more research has been done on the Jewish communities.

Some critics of the DNA screening technology complain that its enthusiastic practitioners promise the world, from early detection and cures for cancer, heart disease, ageing, mortality and everything else, but in fact deliver next to nothing. This is being unfair. The subject is still in its infancy, having only begun to develop properly over the last decade. Already some practical benefits have become apparent. Screening for a common mutation in one of the proteins involved in blood clotting is a good example.

Blood clotting

The prevention of blood clotting has been a real success story. A common mutation has been found where there is clear medical justification for the carrier to know about it. It concerns one of the commonest causes of sudden death in Western countries. A blood clot forms in one of the veins of the lower leg causing tenderness, pain and a swollen calf. The real harm is done if the clot comes adrift and is swept away in the bloodstream where it can often come to rest in one of the large blood vessels supplying the lungs to form what is known as a pulmonary embolus. This can cause sudden death if the blood clot is large enough. The victim asphyxiates. Smaller clots can still cause damage and it is important to prevent their occurrence.

Several conditions favour the development of clots in the calf. If the person is dehydrated, has remained immobile for long periods of time or if there is stagnation of blood in the lower legs for any reason, perhaps because the individual is wearing tight socks or garters, the risk of clotting is greatly increased. Long-haul flights are a common contributory factor: the passenger is cramped in a small seat, cannot move his legs freely, and blood pools in the lower legs. The cabin conditions of the plane favour dehydration making the blood 'thicker' and again more likely to clot.

In addition to these environmental factors, genetic determinants have recently come to light. One of these is a mutation in a gene that codes for a protein (called Factor V) that normally acts as a brake on the clotting process. It keeps the blood fluid and circulating freely. If this brake becomes ineffective, as it does in the mutation (called Factor V Leiden), there is a three- to seven-fold increased risk of developing a blood clot in the legs. The occurrence of this mutation is very high in southern Sweden, with a frequency of about 15 per cent, and this is probably the country where the mutation first occurred. The frequency decreases as you travel further south into Europe. In Germany and the UK the frequency falls to about 5 per cent; it is not found at all in Asia or in sub-Saharan Africa. The mutation may previously have had a survival value that could account for its high frequency in Sweden. It would be expected to help to prevent haemorrhaging after childbirth and so may save the mother's life. Under modern conditions the mutation has become more of a nuisance to women. Thus women who use oral contraception and also have the Factor V Leiden mutation have more than a thirty-fold increased risk of getting blood clots compared to women who are not Factor V carriers and do not use the pill. Clearly, it is beneficial for women who are thinking about using the contraceptive pill to know if they are carrying the mutation.

There are several drugs that can 'thin out' the blood, thus making a clot less likely. The simplest of these is aspirin and some doctors believe that carriers of the mutation should be on long-term treatment with this drug. If carriers experience conditions that make clotting more likely, such as undergoing surgery or undertaking frequent long-haul flights, there are stronger medications that can inhibit clotting. One of these is warfarin and some doctors advise prolonged treatment with this for carriers who have already experienced clotting in their legs or elsewhere. Perhaps everyone should be screened for this mutation if they or their close relations have a history of abnormal clotting.

The future

It is to be expected that many more genetic tests will come to throw light on other common disorders over the next few decades. These will involve screening for large numbers of genetic mutations.

A recent technology, called DNA 'chips', has been developed to look for thousands of genetic variants or markers at a time in any one person. These chips are specially treated glass slides or other solid material, on which are attached DNA strands containing gene coding sequences called probes. These probes match the coding sequence of the genetic variants that might be found in the person's cells. The position of each probe on the chip is exactly established in a chequerboard pattern of microscopic blocks. If the chip DNA meets its matching variant from the person's DNA they will lock tightly together. It is the strength of this binding that permits detection of the person's genetic variants using a clever arrangement of fluorescent tags and optical scanning.

In practice the patient's DNA is extracted from a suitable source of cells, such as the white blood cells, cut into small pieces and then poured on to the DNA chip. Red or green fluorescent dyes are then used to tag the bound DNA of the patient. One can then detect which fragments of the patient's DNA have stuck to the glass slide and from their positions tell which gene variants the person possesses. Such chips can already be used in the analysis of more than 40,000 gene variants that may predispose to cancer, early heart attacks or other common disease. The technique is already streamlined and about 2,000 patients can be studied in one day. When working at full stretch using these new techniques it will be possible for a laboratory to generate mind-boggling amounts of data. It has been estimated that a hundred gigabytes of data could be processed each day in the laboratory. That is about 20,000 times the amount of information in the complete works of Shakespeare or J. S. Bach.

The interpretation of the results of such binding tests is still in its infancy. The final assessment is clearly going to be statistical, which introduces an element of uncertainty into the final outcome. Assigning particular risks for the presence or absence of each of, say, fifty genetic variants and their interactions, as well as accounting for genetic variants that may be protective for development of the condition, will not be an easy task. But it may in the end predict a clinical outcome with a measure of calculated uncertainty. This moves away from the idea of a strict genetic determinism that has held sway for so long from the work on the single gene disorders, such as muscular dystrophy or cystic fibrosis. But even single gene disorders, such as the inherited defects of the blood pigment haemoglobin, can show considerable variability when it comes to the disease they produce. This is especially so if there are interactions with complex environmental factors such as infections with parasites.

Eventually it is hoped that the 'chip' technology may be used to study the variation of all 35,000 or so genes in the human cell so as to obtain a complete genetic profile of the individual. If the cell comes from an early embryo then one will know the future genetic potential of the child to be. A one-inch-square chip has already been developed to hold more than 300,000 independent DNA probes. This is enough to screen for many different genetic variants at each of the genes in a cell. The capacity of future DNA chips will surely increase and probably double every eighteen months if experience with the rate of growth of computer chips is anything to go by. Using such chips the complete genetic profiles could be obtained for a number of individuals or embryos after a few hours' work. It is this technology that will be of critical importance for the creation of 'designer' babies, as described in Chapter 3.

13

Personality Traits

I punish children for the sins of their fathers
to the third and fourth generations.

—Deuteronomy 5:9–10

One popular law of heredity states that the undesirable personality traits of your children never come from you but from your partner. If this were true it would suggest that one's personality is inherited. Plato believed this. In the *Republic*, he imagined the inherited component of character to be made up of different types of metals. People either had streaks of gold, silver, bronze or iron in their personalities (some could also have mixtures of these). People with gold were the most valuable to society and destined to become the elite class of rulers and guardians. People with bronze or iron were marked out to become manual workers, such as farmers or artisans. Generally, fathers with gold would tend to have children with gold; silver with silver; and so on. Plato admitted this idea was what he called a 'noble lie' but hoped it would be adopted because he thought it would provide order and stability amongst the people and keep everyone happy in their own social class. This idea still persists when we say someone is showing their 'mettle', an old variant of the word 'metal'.

Common sense tells us that there is likely to be an inherited component to personality – but to what extent? Again studies of identical twins reared apart (not sharing a common home life), compared to non-identical twins, have given estimates as to the extent of inheritance of personality traits. Identical twin pairs when compared to non-identical twin pairs are found to be much more similar over a whole range of personality features such as intelligence tests, choice of jobs, choice of favourite hobbies and sports, even down to preferential use of adjectives and adverbs.

Of course some identical twins, despite having exactly the same genes, can be very different in personality. But intrauterine factors such as access to the placental blood supply or problems at birth of the second twin may account for this. Mean heritability values of about 40 to 60 per cent have been found to account for the genetic effects on personality traits. In other words about half of our personality characteristics are determined by our genes, and the other half by environmental factors such as our education and other life experiences.

Some forms of behaviour seem to oscillate between being called a disease at one period in time or a normal habit in another. In the nineteenth century masturbation was thought to be a disease that would destroy both health and spirits, leading to feebleness in body and mind and often ending in a lunatic asylum (in the opinion of Lord Baden-Powell). In the twentieth century it became a cure for sexual frustration – or, as Woody Allen put it, like having sex with someone you love.

There is a whole group of conditions that lie on the borderline of clinical disease and abnormal personality traits. Excessive consumption of food or alcohol can be driven by psychological factors and lead to obesity and alcoholic liver disease. Excessive display of anger and aggression leading to criminal behaviour can clearly have psychological determinants, but as we shall see later (see page 242) may also have a possible genetic basis in rare instances. It is this grey area between disease and personality traits that is going to give the most trouble in the future. When dealing with established disease the idea of eugenics seems perfectly respectable to the majority of people. When it comes to personality traits there is real cause for concern. This is where the spectre of Nazi genetics raises its head again.

Obesity

Doctors are always working to preserve our health, cooks to destroy it. But the latter are more often successful. Mounds of marshmallow flesh and breasts like cottage loaves hanging in knitted sweaters are increasingly found all over Western Europe and the United States. Such is the extent to which demand for the diseases of Western affluence exceeds the supply.

People often complain that their obesity is due to their glands – if so, it is most commonly their salivary glands. By far the commonest cause of obesity is the excessive consumption of food – the minor sin of gluttony. Eating food is stimulated by a variety of psychological factors such as a sense of hunger, and terminated by a sense of satiety. Other psychological factors can override the sense of satiety and lead to overeating. In some people all types of stress, such as conflict at work or domestic upsets, can lead to comfort eating whereby the sense of stress is diminished. In turn this overeating can lead to feelings of guilt and thereby accentuate the stress. This inappropriate behaviour pattern does not tackle the main cause of the stress; the individual only responds in one way, namely by eating more food. The situation is made worse in the sense that he or she becomes obese and also leaves the lifestyle problems unresolved. Other types of inappropriate feeding behaviour are eating out of a sense of boredom or overeating as a result of bad habits learned as a child.

Obesity is a common and serious disorder responsible for much ill health. This is partly due to the other diseases to which it contributes, such as diabetes mellitus, raised blood fats, heart attacks and wear and tear of the joints producing a painful arthritis. Obesity occurs when the food intake exceeds the energy output, and the surplus calories are laid down as fat. Some conditions alter energy expenditure, particularly lack of physical exercise, and many doctors continuously proclaim the value of exercise.

But four out of five patients that I see in clinics are more in need of release from anxiety and stress than exercise. However, obesity has been attributed to another of the seven deadly sins, namely sloth. There are also medical conditions such as being short of thyroid hormones that cause a reduction in energy expenditure and are often associated with weight gain.

Can obesity ever be due to faulty genes rather than mere sloth or gluttony? The clue comes from inherited forms of obesity in rodents such as mice or rats that clearly demonstrate that something in the genes can make them fat. Now the same thing has been shown in humans. It is very rare, occurring probably fewer than one in 50–100,000 births, but several children have been found who have defects in the genes that regulate their appetite. The children go on eating without ever satisfying their hunger and become grossly obese. Although the condition is very rare it does give insight into how the normal appetite is regulated. Minor genetic defects may affect the production of neurological signals that regulate appetite and may well occur in people who find it difficult to exert proper control over their appetite. Knowing the pathways involved in the regulation of appetite may suggest new lines of treatment that influence particular steps in the process and so one may be able to suppress the excessive food intake by pharmacological means.

Alcohol (and other addictions)

Alcoholism is one of those terms that are difficult to define precisely. An alcoholic has been defined as someone who drinks twice as much again as you do – so you actually never become an alcoholic! To others it is when your vascular system is running on gin or whisky. But attributing alcoholism to alcohol is hardly more illuminating than ascribing being overweight to overeating. This illustrates the difficulty. To do genetic testing properly you need a precise definition of the condition so that you can accurately identify affected people to study. Admittedly if definitions of such concepts were really possible no one would need to make them. Most definitions end in ambiguity.

With alcoholism there is a variable dependence on drinking. As Mark Twain said: 'It is one of the easiest things to give up. I should know, I've done it a hundred times.' Another time he complained that he was so upset reading about the harmful effects of alcohol and smoking that he decided to give up reading. For others giving up drinking will not make them live longer; it will just make it seem longer. The condition is not like a cancer or a heart attack where you can see or show up the disorder in concrete terms. The damaging effects of alcohol on the liver and brain are very variable. Some people can drink one and a half bottles of wine daily with no obvious ill effects. Others suffer damage when consuming a quarter of this amount.

The causes of alcoholism are also very varied. There are social factors: undue stress at work can induce a person to start drinking for the feeling of pleasure and relaxation that it produces. First you take a drink, and then ultimately the drink takes you. The habit takes hold and because of increasing tolerance to the psychological effects of alcohol more is consumed, dependence is established, and damage to liver and brain can result. Alcoholic liver disease is one of the commonest causes of liver failure in Europe and North America.

Are there genetic factors? Several studies have shown conclusively that severe alcoholism can be predisposed by genetic factors that may account for about 50 per cent of the liability. A typical study of male twins involved 861 pairs of identical compared to 653 non-identical twin pairs. A structured personal interview was used to define alcohol abuse and dependence. This was a weak part of the study because many people have a tendency to misreport or to exaggerate their actions when responding to such questionnaires. The conclusion was that up to 58 per cent of the liability to abuse of alcohol came from genetic factors, the remainder being attributed to environmental influences.

On the other hand there are some purely genetic factors that can prevent people from drinking alcohol. Some oriental groups in particular have an inherited defect in an enzyme that fails to break down alcohol properly in their body. After drinking very small amounts of alcohol they tend to develop uncomfortable symptoms such as facial flushing, headaches, dizziness and sweats. This is sufficient to discourage them from drinking alcohol any more.

Which genes might be involved? The first evidence came from genes that make some of the catalysts (enzymes) that break down alcohol in the body (alcohol dehydrogenase, acetaldehyde dehydrogenase and cytochrome P450, to name a few). Particular variants of these genes were found more frequently in alcoholics than in control subjects. Then a group of genes were implicated that could affect personality traits such as one's pleasurable response to drinking, the so-called 'reward centre genes'.

'Reward centres' are nerve structures deep in the brain that, when stimulated, give a rush of pleasure. The sensation can be like having an intense orgasm for hours on end. It is thought that many things such as alcohol, nicotine, cocaine, heroin and other dubious substances become addictive in some but not all individuals because they overstimulate these 'reward centres'. The addictive substance triggers the release of neurotransmitters such as dopamine or serotonin, which activates the 'reward centre' giving you an intense feeling of wellbeing. Hence the intense craving for more.

Is drug addiction a disease or just an example of a lack of moral fibre in the ability to control the craving? Genetic variants of these 'reward genes' involved in the handling of dopamine were found at greater frequencies in alcoholics than in non-alcoholics. It is a plausible hypothesis that such variants may contribute to the liability of affected people to become addicted. Moreover other studies have used a battery of genetic markers (more than 290 of them) that cover all the chromosomes to scan the genes in 830 alcoholic subjects. A gene located on chromosome 16 was found to associate strongly with the disease. The chance of this observation being due to random effects was less than one in 4,000. What this means is that if you toss a perfectly made coin 4,000 times you would expect to get heads half the time (i.e. 2,000) if it was only down to chance. In fact the study only showed the equivalent of one head out of 4,000 tosses, suggesting some sort of causal factor was skewing the effects of chance against turning up heads.

So one is led to the conclusion that genes may indeed play a part in the development of alcoholism despite the disease ultimately being the result of a complex interplay of other factors. Reproducible evidence that only one or two major genes are involved has been scanty. This is no doubt due to the difficulty of defining the condition in the first place and the fact that it is such a mixed group of disorders.

Evidence that genetic factors can determine a preference for alcohol also comes from studies of laboratory animals such as mice. There is one type of mouse, the C57 strain, when given the choice of either drinking dilute alcohol or plain water will always go for the alcohol. The mouse has not been exposed to undue stress at work, had an alcoholic father-mouse, or an unhappy nest life, but just prefers to drink alcohol. Even to the point of getting plastered.

Intelligence

'I am an intellectual chap and think of things that would astonish you,' wrote W. S. Gilbert. Intelligence is one of those words that is easier to use than define. One eminent psychologist, appropriately called Dr E. Boring, wrote in 1923 that: 'All we know about intelligence is that it is what the test tests', a circular definition if ever there was one. A more usual definition might be a superior ability to solve problems; but this begs the question of which problems are to be solved.

There may be at least three different approaches to solving problems: analytical, practical or creative. Clearly very different mental attributes are required to solve a mathematical problem such as Fermat's last theorem as opposed to the intelligence needed to solve the managerial problems of two colleagues in serious conflict in the workplace, or how to survive as a homeless child on the streets of Rio de Janeiro. In all problem-solving memory, reasoning and thinking ability to conceptualise the issues, imagination to try new approaches, and the ability to learn from past mistakes are going to be involved in varying degrees.

Too much has already been spoken about and published on the thorny question of whether intelligence is inherited or acquired by education and other social influences. Francis Galton was perhaps the first to tackle the problem using statistical methods to show that natural abilities such as intelligence may be inherited. His ideas came from the study of the families of eminent men of science, literature, music and the arts. He found many more than the expected number of gifted members in such families. One has only to think of the Haldane, Huxley or Darwin families to see his point. He published his results in his book *Hereditary Genius* in 1892, but his methods really did not allow him to distinguish between nature and nurture with regard to the acquisition of intelligence.

In 1905 Alfred Binet published the first intelligence tests. This had nothing to do with trying to prove the inherited nature of intelligence or even to measure intelligence. It was simply to sort out a practical problem of identifying children who might not benefit from the usual education provided by public schools in Paris at that time. The tests were quickly taken up by English and American psychologists and elaborated to include a hotchpotch of problems related to vocabulary usage, classification of objects, numerical manipulations and non-verbal tests such as pattern recognition. By 1909 Cyril Burt, an English psychologist, was using these tests to prove that the Intelligence Quotient (IQ) derived from these tests was primarily inherited. It subsequently transpired that much of his data was faked, and even the names of some of his research assistants were fabricated. This threw the whole field of study on the inherited nature of intelligence into disrepute.

The unsatisfactory methodology has continued to be used until the mid-1990s. The American authors Herrnstein and Murray published their book *The Bell Curve: Intelligence and Class Structure in American Life* in 1994. This contained an inflammatory mix of statements and observations about genetics, race and intelligence.

They maintained that IQ tests measure cognitive ability with sufficient accuracy to compare and rank different ethnic groups; that IQ scores are a scientific measure of what we generally mean by 'intelligence'; that properly administered IQ tests are not demonstrably biased against social, economic, ethnic or racial groups; and that cognitive ability or IQ is a major factor dividing the social classes in America. They argued that low IQ is a stable, inherited factor and as measured in people leads to (but is not caused by) undesirable socioeconomic conditions and behaviour. These include poverty, school truancy, unemployment, illegitimacy, divorce, welfare dependency and higher crime rates. This, they claimed, could warrant the withdrawal of public funding and support programmes such as 'Head Start' for low-IQ families to discourage them from perpetuating their own kind with equally low IQs. The authors then went on to rank racial groups by IQ, with African-Americans coming out lowest, with a mean IQ of 85; then white Americans of northern European descent, averaging an IQ of about 100; still higher were the east Asians; and at the top were the Ashkenazi Jews. Anticipating that they might be accused of racism and prejudice against black people they stated that they 'cannot think of a legitimate argument why any encounter between individual whites and blacks need be affected by the knowledge that an aggregate ethnic difference in measured intelligence is genetic instead of environmental'. This is not the place to give a rigorous scientific evaluation of their ideas – others have done so far more competently than I could (see *The Mismeasure of Man* by Stephen Jay Gould). Suffice it to say that the response to the determinist ideas of their book was overwhelmingly negative. The adverse criticism was mainly based on insufficient attention of the authors to eliminating cultural and temporal bias from the very narrow range of intelligence tests that were used and their inadequate use of statistical analyses to derive their conclusions.

However much the field has been strewn with superficial entanglements by previous workers designed to trip up the reader, there is absolutely no doubt that genes can affect intelligence in individuals. To cite just two examples: there is a well-known disorder called the Fragile X syndrome linked to the female X-chromosome. This is the commonest cause of mental disability in men occurring in about one in 4,000 male births. Men only have one X-chromosome and a genetic defect here causes difficulties in learning. The men usually need protected employment as adults and are unable to lead independent lives, needing some form of institutional care. The same genetic defect can occur in women (who have two X-chromosomes) but in this case it is a much milder condition. This is because they can rely on their unaffected X-chromosome to work properly. They are often not recognised as disabled but just have some minor learning difficulties.

The responsible gene on the X-chromosome produces a protein called FMR-1 whose function is still unknown. The condition is called Fragile X syndrome because a region close to the affected gene can undergo a dramatic structural change over the course of a single generation. The region usually contains a short repeated sequence of up to fifty copies of DNA building blocks (the bases CCG). These occur outside the actual gene but somehow interfere with the gene function. The repeated unit can suddenly expand to two hundred copies by unequal sharing of DNA – hence it is called fragile. The longer DNA sequence blocks the gene from producing its protein properly and leads to the intellectual disability. This condition demonstrates beyond any doubt that intelligence defined by gross changes in learning ability can be affected by a single gene, and the defect can be compensated by a normal gene on the other parental chromosome. The learning disability is much milder in affected women. It will be fascinating to learn what the gene product, the FMR-1 protein, actually does.

Other commoner causes of learning disability are minor chromosomal rearrangements in which small pieces of one chromosome get stuck on to another chromosome. Out of 400 children studied with moderate to severe learning disabilities, about 7 per cent had such subtle abnormalities of their chromosomes. This would make it the second commonest cause of learning disability after Down's syndrome. However, there is no evidence that either of these types of abnormality segregates in any particular ethnic group; nor would they be expected to do so.

Recently a gene has been inserted into the DNA of a mouse to produce an animal with superior learning abilities – the Doogie mouse. This mouse was genetically engineered by scientists at Princeton University and the Massachusetts Institute of Technology in Boston. The work was recently reported in the scientific journal *Nature* (1999). When tested on six learning and memory problems the Doogie mouse consistently performed better than an untreated mouse. Doogie learned more quickly, remembered what it had learned and was better adapted to environmental changes than unmodified animals.

The added gene was a component of a type of receptor that is found in plentiful supply in the nerve cells of the memory region of the mouse's brain (the hippocampus) and whose number drops off drastically as the mouse ages. The receptor is situated at the end of nerve cells and is used to respond to incoming signals from adjacent nerve cells. It has been known for some years that if this receptor is blocked by drugs or destroyed by other means it causes a severe learning disability and can even render the mouse amnesiac.

Doogie was created by inserting extra copies of a DNA component coding for part of this receptor into the genes of an ordinary mouse at the single-cell stage of embryonic development. Then the mouse was allowed to develop into an adult. This technique of genetic enhancement created a mouse who was smarter at avoiding unpleasant stimuli such as mild electric shocks to its paws. Doogie could also remember better where hidden objects were located in its cage, and took a livelier interest than the native mouse in novel objects placed in its cage. The gene is not a unique genetic key to intelligence or even to memory. But the plasticity of the brain to lay down new memory traces probably involves many different molecules, of which this is likely to be just one. If one knew the critical components of how memory traces are formed in the brain it might lead to the development of drugs that, for example, could stimulate this receptor and perhaps improve the memory as one aged. If a drug were to be found as cheap and easy to use as aspirin probably everyone would take it. It may eventually be possible to introduce the gene itself into adult humans to improve our memory in later years, in the way that viral genes are introduced during vaccination to protect against diseases such as smallpox.

Intelligent behaviour depends on a whole variety of brain functions working in a coordinated manner. These include short-term memory, long-term memory, learning, perceptual and motor skills, and there may be batches of genes that affect some of these components. But someone has been astute enough to say that great intelligence is just a form of subtle misfortune. People might argue that we should not expend so much energy in trying to improve our intelligence. This may not be a major issue for our future progress. What really matters, what really make us human, are personality features not measured by IQ tests, such as compassion, sensitivity, affection, altruism, judgement and empathy. It is perhaps these more elusive factors that we should try to understand.

Longevity

Why do we die when we do? Some claim divine intervention masquerading as disease, whilst others just consider it to be disease alone. Originally there were the plagues and other infectious diseases; but now in Western societies death is more likely to be due to one of the chronic degenerative diseases. However, if one could accurately predict life expectancy the insurance industry (as well as one's relatives) would be extremely interested. Underestimating the average life expectancy by just two years could easily cost one of the UK's big life assurers up to £200 million. Also the value of annuities depends on a good prediction of life expectancy, and insurance actuaries have consistently underestimated the rate at which our life spans are increasing. As a result selling annuities has not been a very profitable business – for the insurers that is, not for us.

Average life spans for men vary considerably around the world: seventy-six in Japan, seventy-four in the UK, seventy-two in the USA, dropping to thirty-three in Sierra Leone. Most of these differences are due to the prevalence of diseases such as cancer, heart attacks, strokes, malnutrition and infections.

But the normal ageing process on its own, and therefore life expectancy, does appear to be under some genetic controls as revealed by twin studies. In Sweden 3,656 pairs of identical twins were examined, and in Denmark 2,872 twin pairs, to see if the ages of death of each of the twin pairs were correlated, compared to non-identical twin pairs. Both studies came out in remarkable agreement that genes contribute about 30 per cent to longevity. The other 70 per cent is made up of non-shared individual factors such as the lifestyle habits of smoking, drinking and exposure to accidents.

Hitherto the genetic component of ageing was considered to be due to a progressive accumulation of mutations due to errors in the copying of chromosomes, when more and more cells of the body divide inaccurately over the years for repair or replacement of defective body parts such as the skin, liver or intestinal walls.

However, a more specific defect has recently come to light in the DNA that sits at the ends of the chromosomes, called the telomeres. If one likens a chromosome to two shoelaces tied together by a reef knot, then the telomeres are the small plastic tips at the end of the laces to stop them fraying. The 'plastic tips' at the end of the chromosome stop the ends from 'fraying' when the chromosome is copied. It consists of a meaningless stretch of letters such as TTAGGG repeated over and over again up to as many as two thousand times. But every time the chromosome is copied during cell division a small piece of the telomere is lost, so it gets shorter and shorter with each round of chromosomal copying. On average about thirty letters of the telomeres are lost each year. When it shrinks to almost nothing the ends of the chromosomes become 'frayed' and the cells cease to thrive. The telomeres can be repaired by an enzyme called telomerase, but it is not normally made in the cells of the body. It is only produced in the germ line cells that make either sperm or eggs.

Switching off the manufacture of the repair enzyme telomerase is like setting a stopwatch for senescence of the body's cells. In the germ cells the watch is never started for the obvious reason of the need to provide 'young' cells for the next generation. The loss of telomeres appears to be a major reason why cells grow old and die, but it is obviously not the sole reason why bodies grow old and perish.

In very rare cases, such as in the disease called Werner's syndrome, affected people show signs of premature ageing in their twenties or thirties. They go grey early and then lose their hair; their muscles weaken and waste, and their bones fracture easily at the slightest trauma. They develop heart disease, cataracts and 10 per cent of them get cancer. The mean age of death is in the early forties. The cause appears to be due to an accelerated shortening at the tips of the chromosomes, the telomeres, for reasons unknown, and this in turn impairs the accurate copying of chromosomes leading to a 'caricature' of the ageing process. This genetic defect shows convincingly what can occur to accelerate the ageing process.

Animals such as mice also shorten their telomeres with age; and animals that cannot repair their telomeres with the specific enzyme telomerase show signs of premature ageing. Indeed, mice that are totally deficient in this enzyme can only breed for about six generations due to the decreased viability and fertility of their later offspring, so this may be one genetic factor affecting our life span. It is more likely that longevity is under the control of many other genes. One informed estimate suggested that more than 7,000 genes could be involved in the ageing process. It is therefore nonsense to talk about the 'ageing gene' except in the very rare disorders such as Werner's syndrome. Ageing is due to a simultaneous deterioration of many different body parts and until we understand all the factors involved the old-fashioned advice still applies: don't smoke, drink only in moderation, take regular exercise and have a happy marriage. For a man who is divorced, widowed or lives alone, statistics tell us that he is more likely to die early. For women it is the opposite; if they don't marry they tend to live longer.

Aggression

Violent aggressive behaviour such as assault against someone is usually treated as a criminal offence. It is far more common in men than women, suggesting it may have biological determinants. The perpetrator of the assault, if mentally healthy, is held responsible for the crime. He cannot excuse his behaviour in a court of law on his constitution, his genes, his social background of poverty or emotional deprivation. But now the law may have to be modified, or at least take into account some of the more recent genetic discoveries.

In 1993, a large Dutch family was studied by Dr Brunner and colleagues at the University of Nijmegen, of whom eight out of fourteen members then alive showed a characteristic type of aggressive behaviour that was difficult to diagnose in current psychiatric terms. No women in the family were affected. The eight affected males all had a mild degree of mental retardation with a mean IQ of eighty-five. They displayed repeated episodes of violence usually triggered by anger. Some had molested young women and one had raped his sister. Subsequently, while on a prison farm, he stabbed a warden in the chest with a pitchfork. Another got angry with his boss and tried to run him over with a car. Another entered his sisters' bedroom at night armed with a knife and forced them both to undress. At least two of the men were known arsonists. The disorder showed a classic pattern of sex-linked inheritance, passed on by the women of the family, but showing its effects in the men. The urine of the affected men contained large amounts of certain chemicals involved in brain signalling.

Another surprising finding in this family was a genetic marker on the X-chromosome that tracked with the affected men. The odds against this genetic marker being linked by chance were about one in 1,500. This was eventually traced to a faulty gene in the neighbourhood of the marker that may have predisposed the affected men to behave in such a violent and antisocial manner. The gene (coding for the enzyme monoamine oxidase) is involved in the metabolism of hormones for 'flight and fight'. This perhaps makes sense of the symptoms of anger triggering bouts of aggression. However, not all the men who inherited this bad gene have committed a criminal offence. One of them is married with children and holds a regular job. Perhaps if environmental stresses become too great they are more likely to trigger a burst of rage from the predisposed men. However, it appears to be a very rare syndrome. No other families have so far come to light, but the discovery may throw light on the biochemical details of how people manage to regulate their aggressive social behaviour within defined limits.

It is possible that other psychopathic personalities leading to criminal behaviour may have genetic conditions underlying their behavioural disturbance, in the same way that the disorder of colour blindness can be genetically determined. Both groups just cannot 'see' or appreciate the consequences of their actions. The question as to whether psychopathic men should be rehabilitated in hospital because of a possible unfortunate genetic endowment or punished in prison as criminals is a highly controversial topic. But in either case they do have to be excluded from society for the public's protection.

Homosexuality

Probably one of the most sensational 'discoveries' of the last decade was the announcement by Dean Hamer in the scientific journal *Nature Genetics* (1995) that he had found a gene on the X-chromosome that determined sexual orientation. In a study of thirty-three families comprising 113 individuals in whom at least two brothers were homosexual, a genetic marker was found on the X-chromosome (Xq28) that was associated with male sexual orientation. The odds for this association occurring by chance were less than one in 2,000.

No independent laboratory has yet confirmed this work, but previous studies of twins have suggested that homosexuality is highly heritable. For example, among fifty-six gay men who had identical twins there were twenty-nine whose other twin was also gay. But among fifty-four gay men who had non-identical twins, there were only twelve whose other twin was gay. So it remains a plausible hypothesis that there is a genetic predisposition to some forms of homosexuality. Other forms might be learned or acquired in early adult life.

Informed theological opinion has already taken up the idea to consider screening embryos for the 'gay gene' and then to specu-late as to whether we should introduce abortion as a means of cleansing our society of a 'gay' population. No less a figure than Field Marshal Montgomery, commenting on a bill to relax the laws against homosexuals in 1965, said: 'This sort of thing may be tolerated by the French, but we are British – thank God.' (This is somewhat ironic in view of Monty's official biographer Nigel Hamilton claiming that Montgomery had strong feelings for his young soldiers.) Such public hostility to homosexuality has now largely disappeared in most European-based societies and such comments would meet with the derision they deserve.

Anxiety

It has been said that a psychotic person knows that two plus two makes five and is perfectly happy about it, whereas a person with an anxiety neurosis thinks that two plus two makes four and is ter-ribly worried about it.

Anxiety is a sort of fear spread thinly over all one's daily activ-ities. The feeling is akin to being very hungry and having a dish of delicious food in front of one that one knows to be poisoned. This generates a strong emotional conflict of approach-avoidance and leads to many of the features of anxiety. These traits in an indi-vidual appear to be quite fundamental and enduring aspects of their personality. The level of anxiety can be 'measured' by several different questionnaires such as Cattell's personality inventories.

Personality features such as avoidance of dangerous situations, worry, tension, pessimism, fear of uncertainty and fatiguability all show close correlations. The question now arises as to whether these 'scores' for anxiety neurosis relate to any types of genetic variants. Twin studies have suggested that up to 60 per cent of the individual variation in measurements of anxiety traits of personality can be due to heredity. So it would not be unreasonable to look for genetic associations.

A recent study of a sample of 505 individuals from the general population did indeed show associations with a particular genetic variant (the 5-hydroxytryptamine transporter gene) that codes for a chemical involved in the transmission of signals within the brain. Two types of genetic tests were performed, a population association study and a paired sibling analysis within families. They both showed significant associations between anxiety traits and the genetic marker. Individuals with the short version of the gene had greater anxiety related scores than individuals with the long version of the gene, the difference in length between the two variants being about forty-four bases. The work was published in the highly respected journal *Science* in 1996 and had passed the critical review by scientific peers. But the work cannot be taken too seriously until the observation is repeated and confirmed in different population samples by workers in other laboratories. This confirmation is still lacking.

It is perhaps naively oversimplistic to consider that one gene can determine a major personality trait in a healthy person. Even if the finding is not completely wrong it is very dangerous to oversimplify the genetics of personality traits. In the past others have acted on the basis of such simplistic ideas to produce catastrophic results. The Nazis' breeding programme for the purification of the Aryan race was one such case, but this type of approach will certainly lead to an extensive search for other anxiety related genes perhaps with greater effect than the currently reported one. Knowing that more than two hundred genetic variants could be involved in the development of a simple visible defect, such as the accumulation of a fatty plaque in one of the major arteries, leads one to believe that perhaps many hundreds of genes could be involved in the development of a personality trait such as anxiety.

Information on such genes could be useful in the long run to optimise pharmacological treatment of neuropsychiatric disorders. Drugs such as Prozac that alter some functional aspects of the protein products of gene action are already on the market and are useful in the treatment of anxiety and depression. Newer and better drug therapies may evolve when other genetic determinants are discovered.

Mood disorders (mania and depression)

It is natural for everyone to have swings of mood during their life, from a sense of elation and success at having won the jackpot on the Lottery, to a sense of depression when one realises one has lost the ticket. But in some people – about 1 per cent of Western populations – these swings of mood become excessive, leading to a severe disruption in lifestyle. During periods of elation the person becomes maniacally overactive. He or she is constantly energetic, no sleep is needed and the brain races away in a flight of ideas that often make the speech garbled and incoherent. Grandiose business or other work plans are initiated. The house is remortgaged or the personal pension fund is pilfered to provide finance for a vastly profitable and universal business plan that no one has ever considered before. The victim believes she or he is amazingly attractive to the opposite sex and embarks on a disastrous series of love affairs. No problem is so hard or complex that it cannot be solved successfully when managed by a person of genius and destiny. Their kites fly higher than anyone else's.

Understandably this behaviour often spells financial disaster and a breakdown of previously established relationships, and can lead to a ruined life. The flip side is when the victim crashes into a deep depression where the world looks gloomy and bleak. There is nothing to do but to stay in bed and stop eating – all one's efforts are futile and meaningless. Life is not worth living and the best course of action would be to save up enough sleeping pills to commit suicide. Such people, often at the instigation of relatives, consult a psychiatrist who usually makes the diagnosis of a bipolar affective disorder, otherwise called a manic-depressive psychosis.

Although this is a remarkably complex mood disorder there is an amazingly simple treatment for it. Prescribe for the patient a simple salt that resembles washing soda. Instead of sodium carbonate, give lithium carbonate. Lithium and sodium are close chemical relatives. How it works is not understood at all, but it can wipe out the episodes of mania in many of the sufferers.

The treatment is not without side-effects – patients can develop stomach upsets or a fine tremor of their hands – but it is so simple that it leads one to think that there may be an equally simple unifying cause for this mood disorder. Could there be genetic factors at work? The usual twin story seems to apply again. If you have an identical twin with this mood disturbance you will have about a 60 per cent chance of getting it yourself. Whereas if your non-identical twin develops the disease you only have about a 1 per cent chance of getting it – no different from the risks in the general population. This indicates that there may well be genetic factors underlying the disease. These would be well worth looking for because the current treatment with lithium is useful but not ideal. If we knew the genetic basis for the condition it could lead to new lines of drug therapy either to replace a defective gene product or to switch off a gene that is making too much of its product.

Some investigations have already had partial success in identifying genes at various positions on different chromosomes. In a very large Costa Rican family a stretch of DNA on chromosome 18 was identified that appeared to be linked to the disease in affected family members. However, in other studies this was not confirmed. Segments of DNA on chromosome 4 appeared to be linked in a large Scottish family, and DNA on chromosomes 6, 13 and 15 in a large Amish pedigree from Pennsylvania. The lack of overlap in these findings is disappointing. Finding the same stretch of DNA in three independent studies would have greatly increased confidence in the importance of the results. These studies probably do represent the first faltering steps in the eventual identification of the genes involved in accounting for the heritability of this disease.

If they knew where these genes were located, would parents ever consider their use in the eugenic sense of screening their early embryos after *in vitro* fertilisation and therefore before the pregnancy has started? It would depend in many ways on their previous experience of the disease. If they had several close relatives who had made a disaster of their lives and committed suicide in early or middle age, prospective parents might well fear the consequences of passing on such genes conferring susceptibility to their offspring.

However, minor forms of the disease can be quite attractive. Instead of the blinding mania, the victims show a more likeable temperament in terms of charm, enthusiasm, creative energy and sexuality that may compensate for short periods of depression that can occur. In such a case parents may have no qualms about passing on such genes. It all depends on their perspective. Some exceptionally creative artists appear to have produced their best work when in a slightly hypomanic mood. Vincent van Gogh (1853–90) had alternating bouts of excitement and despair, the latter leading to incarceration in a mental hospital and eventually to his suicide. When active, however, he worked with great speed and intensity, producing more than 800 oil paintings and 700 drawings between 1880 and 1890. He certainly had the symptoms of a manic-depressive illness. One wonders how creative he would have been if treated by present-day therapy.

But should we leave our artists to suffer so that they may produce paintings that we enjoy looking at? Another victim of the disease wrote eloquently about this:

> I have often asked myself whether given the choice, I would choose to have a manic-depressive illness. If lithium were not available to me the answer would be a simple no – and it would be an answer laced with terror. But lithium does work for me, and therefore I suppose I can pose the question. Strangely enough I think I would choose it . . .

Because I honestly believe as a result of it I have felt things more deeply; had experiences more intensely . . . And I think much of this is related to my illness – the intensity it gives to things and the perspectives it forces on me . . .

This was reported by Kay Redfield Jamison in the book *Mood Genes* by Samuel Barondes.

Schizophrenia

Other behavioural disorders such as schizophrenia and childhood autism also appear to have a genetic basis. Schizophrenia is a serious psychiatric disorder in which the victim can suffer from complex delusions, often of a persecutory nature, from hallucinations that can be very frightening and a general breakdown in thinking processes. Remarkably cannabis, derived from the plant *Cannabis sativa*, can produce very similar hallucinations and delusions in some long-standing addicts. This suggests that there may be a simple chemical problem underlying schizophrenia that can be mimicked by the plant alkaloids found in cannabis. The child of a schizophrenic parent has about ten times the risk of developing the disorder compared to a member of the general population where the lifetime risk is about 1 per cent. When both parents are affected the risk becomes forty-fold greater than in the general population.

The argument that this is all due to social and environmental factors is countered by two lines of study. If one of a pair of identical twins has schizophrenia there is about a 70 per cent chance that the other will develop the disease, whereas for non-identical twins the value is reduced to about 30 per cent.

One could still argue that this does not entirely exclude environmental factors because identical twins report greater social similarities such as choice of clothes and the sharing of friends than non-identical twins. However, a recent study recruited forty-seven children with a schizophrenic mother who were adopted within seventy-two hours of birth. They were later compared in their mid-thirties with fifty control adoptees who did not have a schizophrenic parent. Five of the adoptees (11 per cent) with affected mothers became schizophrenic themselves (roughly the rate expected in non-adopted offspring of schizophrenics), compared with none of the controls. This type of study has been confirmed on several other occasions.

As with other common complex conditions there are probably a multitude of genes involved in psychiatric conditions and this will make it more difficult for population screening to identify people at risk. The knowledge is more likely to be of use in screening to identify high-risk relatives of members of affected families. This could have obvious benefits in advising relatives to avoid environmental risk factors such as the recreational use of cannabis or amphetamines (such as speed or ecstasy) in the case of schizophrenia.

Future issues

The trend in recent years has been to move away from chemical markers in the blood or urine towards the use of genetic markers. These were first used for the rare single gene disorders such as cystic fibrosis or muscular dystrophy and then extended to the commoner multiple gene disorders such as heart disease, diabetes or the dementias. Attempts are now being made to use DNA tests to identify people who have markers for various personality traits that lead to such conditions as obesity, alcoholism, aggressive behaviour or sexual preferences.

Some of these genetic markers have already proved of medical use as aids to diagnosis and to prognosis. A genetic marker in one of the genes coding for a fat-transport protein in the blood can assist in the diagnosis of the dementia of late-onset Alzheimer's disease. Markers can also be used in predicting the risks of an otherwise healthy person who might go on to develop a disease in the future. A good example of this is the genetic variation of the cell surface protein that removes cholesterol from the bloodstream. If an abnormal variant is found, this can predict the occurrence of heart attacks in the future and early avoidance action can be taken. Once a genetic defect has been incriminated for the cause of a disease it can open up new lines of treatment either by gene replacement or by suggesting new drugs that may be able to reverse the defect.

The knowledge and use of genetic markers in the child or adult may give us the ability to predict disease and longevity. When used at the early embryonic stage it may give us the ability to make 'designer' babies. The greater these powers, the more dangerous the potential abuses. Society's response to this development has ranged over a broad spectrum of ideologies. On the one hand, Pope John Paul II pronounced: 'Genetic screening is gravely opposed to the moral law when it is done with the thought of possibly inducing an abortion. A diagnosis which shows the existence of a malformation or a hereditary illness must not be equivalent to a death sentence.' At the other extreme is China's law on 'Maternal and Infant Health Care' of 1995 which makes prenatal testing compulsory, to be followed by termination of the pregnancy if a foetal disorder is found.

14

Conclusions

This World is not Conclusion.
A Species stands beyond –
Invisible as Music –
But positive as Sound –
It beckons and it baffles –
Philosophy – don't know –
And through a Riddle, at the last –
Sagacity must go –

—Emily Dickinson, untitled poem, 1862

Emily Dickinson found it difficult to draw any definite conclusions about the world. It appears just as difficult to draw conclusions about the narrower topic of the new eugenics. However, there are no shortages of opinions and passionate convictions; 'much gesture from the pulpit', as Miss Dickinson goes on to write later.

Still I think there are one or two things worth saying now.

It is worthwhile rehabilitating the idea of eugenics and restoring its original meaning as defined by Francis Galton. In his book *Remaking Eden*, Lee Silver has invented a new term – 'reprogenetics' – to cover the subject matter. This certainly gives due regard to the twin foundations of the subject in genetics and assisted reproductive techniques (to include *in vitro* fertilisation, egg and embryo storage, embryo selection, etc.). But 'reprogenetics' raises none of the historical perspectives. The history of the subject is so important because it shows us in no uncertain light the many horrendous pits that we should try never to fall into again. These have been mainly dug by our politicians and regulators.

The new genetic and reproductive sciences cry out for applications. Indeed, a final test of the validity of any new science is whether it can be used successfully to change the world in any practical way.

Assuming the new methods are safe enough to be applied in the near future raises the difficult issue of who should control them and how they should be regulated.

In the past, the mania for regulation and control has been responsible for the grossest abuses; the actions of certain right-wing politicians in America, Germany and Scandinavia in the early twentieth century have already been discussed in this book. China's policy of enforcing one child per family brings this up to date, a policy which is already altering the sex ratio of newborn children in favour of boys. None can predict how this experiment will affect Chinese society and the population in the future but it is expected that the number of men will far exceed that of women.

In the past the attempts of many Western countries to control the simplest of eugenic techniques, such as abortion, have been bungled. As stated earlier, in the UK the law is absurdly contradictory. Abortion is illegal under most circumstances in Northern Ireland; it is almost freely available in England. If criminalised, abortion may be driven underground, providing an unsafe or even fatal service for women. If legalised, strong and often violent opposition groups may be formed. In the USA, twenty-one doctors were attacked and seven murdered between 1991 and 1997 for practising abortion. In the same period there were eighty-six episodes of bombing and arson against those who were prepared to carry out abortions.

It may ultimately be better to allow individuals to decide for themselves as to whether or not abortion is acceptable; it becomes a matter for personal conscience, something to be judged on a case-by-case basis. This may be the way to manage all the newer eugenic techniques. The thought of embryo selection or storage may indeed appal you, but this is not a good enough reason why another person should not be allowed to use it.

Regulation may be employed more to protect the public from undue or dishonest commercial exploitation. There are signs of such abuse already: genetic tests to identify a predisposition to cancers of the breast and colon are being applied incorrectly in the marketplace.

This raises the general issue of how the new technology should be funded. Should it be left solely to the commercial sector? Already about 80 per cent of assisted reproductive methods in the UK are paid for privately. Costs of *in vitro* fertilisation can run into thousands of pounds. Adding the genetic techniques will make it even more costly. It would be quite iniquitous to confine access to the new technology to a wealthy minority. The methods should be made freely available for anyone wishing to use them. Some, such as *in vitro* fertilisation, can already be prescribed on the National Health Service in the UK. Possibly a specialised insurance scheme could be devised for interested parties to provide funding for the new technology.

A strong groundswell of opposition against the use of the new techniques has recently arisen from many different quarters (including conservationists, theologians, ethicists, and environmentalists). In many ways the geneticists have brought this down upon themselves by promoting overblown statements in the media that favour the case for genetic determinism. We all know that human development results from an intricate interplay of genetic and environmental influences. The nature–nurture antithesis is well and truly dead. It should be consigned to the waste bin together with all the other moribund ideas such as pangenesis, phlogiston and the ether.

Society as a whole should perhaps embrace the new knowledge and the opportunities it offers less timorously, or even with some measure of enthusiasm. The new eugenic technology may become a vital weapon to prevent a future genetic deterioration of our species. Other animals have suffered such a fate. The African cheetah, for example, is a species which faces extinction because of a genetic decline, which involves a loss of diversity in its gene pool. Whatever the reasons for this, the reduction of genetic variation within the species makes the cheetah less able to adapt to environmental changes. Some evidence shows that this animal is more prone to infections because of a poorly adaptable immune system.

Our own genetic decline may take a different form. One hundred years ago some people would never have been able to reproduce. People with early-onset diabetes, premature heart attacks, malignant high blood pressure, or cancers such as the leukaemias, may have survived into their reproductive period, but were too unhealthy to have children. Nowadays improvements in medical and surgical treatments allow such people to lead an almost normal reproductive life. Before the discovery and use of insulin, early-onset diabetics had almost no chance of having children of their own. Medical therapy has restored their fertility, although some diabetic women do still have problems at childbirth.

The consequences of these medical advances are that parents can more freely transmit their disease-related genes to their children. These defective genes would be expected to accumulate from generation to generation in ever increasing numbers.

To prevent this we may need in the future to screen embryos for disease-related genes and if possible to repair them at an early stage using the most powerful tools we have. The new eugenic technology may play a prominent role in this.

Meanwhile, I am sure many of the issues raised in this book will continue to generate heated discussion. Doubts will vie with certainties to influence our future practice. Let the last words go to Emily Dickinson, closing the quotation at the beginning of this chapter as eloquently as she opened it:

> To guess it [the right way] puzzles scholars –
> To gain it, Men have borne
> Contempt of Generation
> And Crucifixion shown –
> Much gesture from the Pulpit
> Strong Hallelujahs roll –
> Narcotics cannot still the Tooth
> That nibbles at the soul.

Glossary

Adverse selection An insurance company term indicating that some people at high risk of disease purchase large policies and stand to make a financial gain out of the premiums paid by the lower-risk individuals. This would in effect be robbing the common pool of people of the money they pay as premiums.

AIDS (Acquired ImmunoDeficiency Syndrome) Destruction of the immune system by infection with a virus called HIV, leading to multiple and life-threatening infections.

Alzheimer's disease A chronic neuro-degenerative disease of the brain often starting with memory loss and leading to a widespread loss of mental function. Some forms of the disease show a strongly inherited tendency.

Amino acids Small molecules that form the building blocks to make proteins. There are twenty different amino acids and their order determines the individual properties of each protein.

Atheroma A fatty-like material that deposits in arterial walls. If sufficiently large to obstruct blood flow it can cause heart attacks and strokes.

Bases In this context, a shorthand word for the four different building blocks (also called nucleotides) that make up the long thread-like molecule of DNA.

BRCA1, BRCA2 Two genes found on chromosomes 17 and 13 respectively whose mutations can predispose a woman to develop breast cancer.

CAT scan A complex X-ray that provides pictures of the body's internal organs.

Chips In this context, a small piece of glass or other material to which is attached a great number of DNA sequences used for the analysis of the variation in an individual's genes.

Cholesterol One of the body's fats that commonly accumulates in arterial walls to form plaques (see **Atheroma**).

Chromosomes Packages of DNA wound into spindular forms and found in the nucleus of cells. Humans have twenty-three pairs of chromosomes, one set from each parent. They contain the majority of the genes for that person.

Cloning A reproductive technique where genetically identical copies of a person are made. Identical twins are clones. Recently it has come to include the reproductive technique of nuclear transfer of an adult nucleus into an empty egg cell to start off foetal development.

Concordance rate In twin studies it is the proportion of twin pairs that show the same characteristics that are under study, i.e. presence of a particular disease. A concordance rate of 100 per cent means that all twin pairs show the same feature.

Correlational analysis A statistical technique to measure the extent of causality between two phenomena.

Cystic fibrosis An inherited disease (affecting about 1:2,000 live births in the UK) that produces sticky mucus secretions in the lungs, liver, pancreas and other organs. A major feature of cystic fibrosis is recurrent lung infections.

Diabetes mellitus A common disease (afflicting as many as 1:10 people) in which the regulation of the blood sugar is impaired to produce abnormally high levels. There are many different types of diabetes, some of which can show a strongly inherited tendency.

DNA Short for **deoxyribonucleic acid**, a long thread-like molecule containing four different bases (called nucleotides). The molecule contains the coding sequence of the genes.

Double helix Two strands of DNA associate together to form a structure like a spiral staircase.

Down's syndrome (formerly known as mongolism) A disease in which an individual inherits an extra chromosome, usually chromosome 21. This gives rise to a multitude of abnormalities including mental impairment.

Duchenne muscular dystrophy A muscle-wasting disease inherited on the female X-chromosome and affecting boys.

Enzymes Proteins that have the power to catalyse (or accelerate) chemical reactions.

Eugenics Defined by Sir Francis Galton as 'the science of improving the inherited stock, not only by judicious and selective matings, but by all other influences'. This would now include all the latest scientific advances in genetic and reproductive technologies.

Factor V Leiden A mutation in one of the proteins involved in blood coagulation making it more likely for a clot to develop.

Familial hypercholesterolaemia A fairly common inherited disorder (affecting about 1:500 people) in which the body cannot regulate the blood cholesterol properly. High blood levels of cholesterol occur and deposit in arterial walls (see **Atheroma**).

Foetus Another name for the developing embryo after about eight weeks.

Fragile X syndrome An inherited disorder where abnormal repeated sequences of DNA occur on the X-chromosome and cause mental retardation.

Gene product Genes make a molecule called messenger RNA whose function is to carry the information coded by the gene to the sites of protein synthesis in the cell.

Genes The fundamental units of heredity consisting of stretches of DNA containing a code from which the cell can manufacture particular proteins. Most genes are found in the nucleus, but some occur elsewhere (for example, in the mitochondria).

Genetic marker A variation in a gene sequence that allows one to track the gene to see to whom it is transmitted or how it is distributed in populations.

Genome All the genes together with the rest of the DNA of chromosomes found in an individual.

Growth hormone A protein messenger secreted by the pituitary gland that stimulates growth.

Haemophilia An inherited disorder of the blood-clotting mechanism where a protein factor is defective or missing. The affected person bleeds excessively from the slightest injury.

Heritability values The proportion of the variation of a bodily feature that can be attributed to genetic as opposed to environmental factors.

HIV Human Immunodeficiency Virus that produces the disease AIDS.

Huntington's disease A rare genetic disease usually starting in adult life, characterised by abnormal limb movements and a progressive dementia.

Identical twins Twins with an identical genetic make-up, usually due to splitting of the embryo into two at a very early stage of development.

IVF (*in vitro* fertilisation) When conception (i.e. the fusion of egg and sperm) occurs outside the body in artificial containers.

Messenger RNA A major product of gene action that carries the information coded in the gene to sites of protein synthesis in the cell.

MRI scans Magnetic resonance imaging scans which use strong magnetic fields to produce very detailed pictures of the human body.

Muscular dystrophy A group of diseases associated with muscle wasting. Some are inherited as single gene defects (see **Duchenne muscular dystrophy**).

Mutation An alteration in the DNA sequence of a gene that may have adverse effects on its ability to code for protein synthesis.

Natural selection One of Charles Darwin's key evolutionary ideas: animals (and plants) that have characteristics most favourable for survival will contribute more offspring to the next generation than others. The most favoured for survival will be those naturally selected, being better adapted to their environment.

Non-identical twins Twins who do not share the same genetic make-up. Two separate eggs are fertilised by different sperm and then both implant in the womb for development in the same pregnancy.

Nucleotides The basic building blocks that make up DNA. There are four: adenine, thymine, guanine and cytosine.

Pangenesis One of Darwin's theories to account for inheritance based on particles called gemmules (a theory long since discredited).

Phenylketonuria An uncommon inherited disease (about 1:10,000 individuals is affected in the UK) due to a single gene defect on chromosome 12 that allows abnormal chemicals to build up in the blood and cause mental retardation.

Proteins Large molecules composed of thousands of amino acids. The exact order of the amino acids determines the properties of the protein that can have many different functions: catalysts, messengers, pumps, gate-keepers, membranes, etc. Proteins comprise the basic working machinery of the cell.

Retinitis pigmentosa A group of degenerative disorders at the back of the eye that can cause blindness. Many types are genetically determined.

Ribosomes Particles found in cells that are the sites of manufacture of proteins directed by the messenger RNA.

Schizophrenia A psychiatric disorder occurring in about 1 per cent of the population and characterised by hallucinations, delusions and thought disorders. There is an inherited predisposition to the disease.

Sickle cell anaemia A genetic defect of haemoglobin that causes a severe anaemia when two doses of the bad gene are inherited (one from each parent) but protects against malaria if only one defective gene is inherited.

Stem cells Embryonic cells that have the potential to form different organs such as the skin, liver or nervous tissue.

Tay-Sachs disease A rare fat-storage disease of the nervous system caused by a genetic defect on chromosome 15. It leads to progressive mental deterioration and usually to death within two years of birth.

Telomeres The tips of the DNA strands found in chromosomes.

Therapeutic cloning The use of cloning techniques to create an embryo from which cells can be harvested for the treatment of various adult disorders, provided the embryo has not gone beyond the fourteenth day of its development.

Transfer RNA A class of RNA that helps to assemble the amino acids in an order determined by the messenger RNA.

Triplets In this context, three nucleotides in sequence that can code for one particular amino acid.

Utilitarianism A branch of philosophy that judges the moral worth of actions by the usefulness of their consequences.

Warfarin A chemical used as an anticoagulant for humans but as a poison for rats.

Werner's syndrome A very rare genetic disorder associated with accelerated and premature features of ageing in late childhood and early adolescence.

Appendix

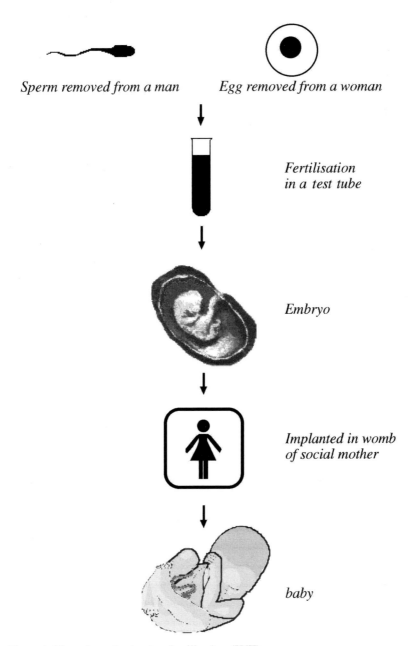

Sperm removed from a man   *Egg removed from a woman*

*Fertilisation
in a test tube*

Embryo

*Implanted in womb
of social mother*

baby

Figure 1. Flow sheet for *in vitro* fertilisation (IVF).

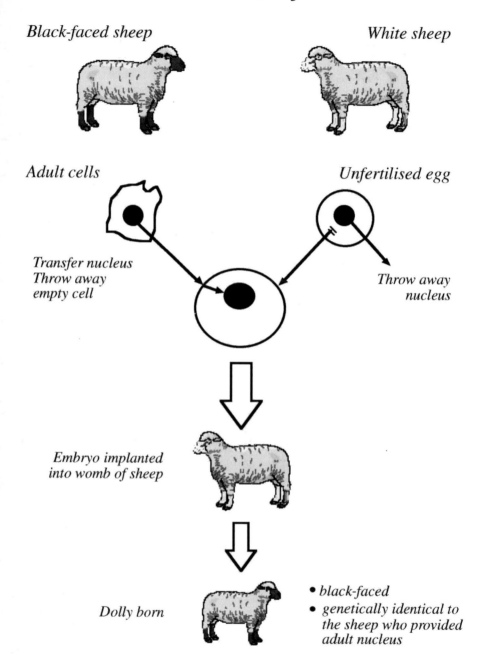

Figure 2. Flow sheet for cloning from an adult nucleus to create 'Dolly' the lamb.

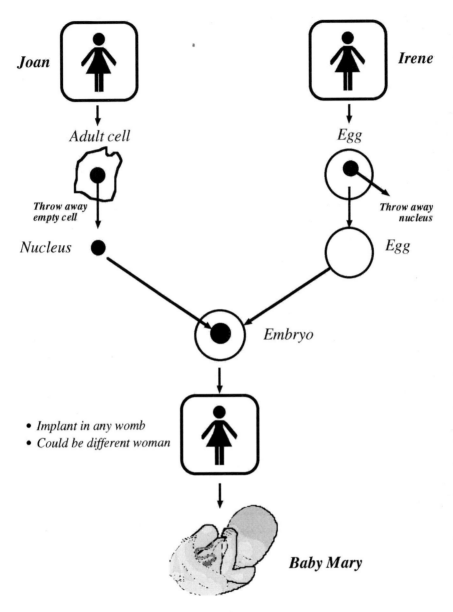

Figure 3. Flow sheet for cloning from two women.

Table 1. Some genetic markers for diagnosis, prognosis or prediction of common disease

Disease	*Frequency	Genetic Markers
Metabolic		
Diabetes Mellitus (Type I)	1:5000	HLA-DR3/4; INS.vntr
Diabetes Mellitur (Type II)	1–7%	INS-R; GLU-t; GK; HNF1a; HNF4a
High blood cholesterol	1–5%	LDL-R; apo B; MTP (-493T)
Cancer		
Breast	25,000 new cases per annum	BRCA1; BRCA2; Her2/neu
Colorectal	20,000 deaths per annum	PMS1; PMS2; MLH1; MSH2
Neurology		
Late-onset Alzheimer's disease	10% of over-65s	Apo E4; APP; S1; S2
Narcolepsy	1:2000	HLA-DR2
Schizophrenia	1:100	Dopamine D2; telomeric X-chromosome
Cardiovascular		
Premature heart disease	1:5 male adults	Apo E and others
Venous clotting	0.1–1%	Factor V Leiden, Prothrombin (G20210A)
Hypertension	15–20%	ACE; angiotensinogen; renin
Rheumatology		
Ankylosing spondylitis	0.5%	HLA-B27

Abbreviations: INS = insulin gene; vntr = variable number tandem repeats; INS-R = insulin receptor; GLU-t = glucose transporter; GK = glucokinase; HNF = hepatic nuclear transcription factor; LDL-R = low-density lipoprotein receptor; MTP = microsomal triglyceride transfer protein; apo E = apolipoprotein E; ACE = angiotensin converting enzyme; BRCA = breast cancer gene; p53 = tumour suppressor gene; APP = amyloid precursor protein; LPL = lipoprotein lipase; PMS1/2 and MLH1 = mismatch repair genes; S1/2 = presenilin genes.

*Frequency refers to either incidence, prevalence, or frequency depending on population studied.

Table 2. **Principles on which to base legislation for the application of predictive genetic testing taken from the Convention on Human Rights and Biomedicine (Council of Europe 1997)**

General:
1. Guarantee individual rights and fundamental freedoms of all individuals.
2. Welfare of individual to prevail over sole interest of society.
3. Equitable access to genetic tests and consequences for all.
4. Any intervention to be carried out in accordance with relevant professional and accreditation standards.

Consent:
1. Any intervention requires free and informed consent.
2. Appropriate information as to the purpose and nature of the intervention must be given.
3. Freedom to withdraw consent.
4. For a minor, need consent of the guardian.

Privacy:
1. Rights to privacy of information about health status.
2. Rights for access of individual to their health information that has been collected (also rights not to know if so wished).

Genetics:
1. Any form of discrimination against a person on the grounds of genetics to be proscribed.
2. Predictive genetic tests may be only performed for health purposes or for scientific research linked to health purposes.
3. Any genetic intervention only to be undertaken for predictive, preventive, diagnostic or therapeutic purposes related to clinical disease.
4. Use of screening techniques during assisted procreation would not be allowed for purposes of choosing a future child's sex except when serious hereditary or sex-related disease is to be avoided.

Sanctions:
1. Appropriate sanctions (removal of accreditation, removal of licence to practise, fines) shall be applied in the event of any infringement.
2. A person suffering damage as a result of such procedures is entitled to fair compensation according to the processes of law.

Table 3. **Examples of some single gene disorders**

Disease	Disease Incidence	Features	Screening Test
Phenylketonuria (PKU)	1 in 10–15,000	Intellectual impairment	Abnormal blood chemical or genetic test
Haemophilia	1 in 5–10,000 males	Bleeding tendency	Abnormal clotting problem or genetic test
Sickle cell disease	1 in 600 (African-Americans)	Vessel clots	Abnormal haemoglobin or genetic test
Familial hypercholesterolaemia	1 in 500	Early heart attacks	High blood cholesterol or genetic test
Cystic fibrosis	1 in 2,000	Sticky secretions; lung infections	Genetic test

Table 4. Heritability for some common diseases

| Trait or Disease | Concordance rates in twins: | | Heritability |
	Identical	Non-identical	(as per cent)
Alcoholism	>0.60	<0.30	60
High blood pressure	0.55	0.25	60
Obesity (body mass index)	0.95	0.53	84
Diabetes mellitus (Type II)	>0.75	0.30	75
Heart Attacks	0.44	0.14	65
Schizophrenia	0.47	0.12	70

The data were compiled from a large variety of sources, primarily European and US populations. A heritability of 100 per cent means that the disease appears to be determined wholly by genes; a heritability of 0 per cent means that genes probably play no part and that the environment is the major cause.

Sources

Seven specialist papers published in medical journals by co-authors and myself give fuller details of many of the original journal sources. These papers are:

Galton, D. J. 'Anton Chekhov's MD thesis: Sakhalin Island'. *Journal of the Royal College of Physicians, London* 23: 189–194, 1989.

Galton, D. J. 'Chekhov after Sakhalin Island'. *Journal of Medical Biography* 4: 102–108, 1996.

Galton, D. J. 'Greek theories on eugenics'. *Journal of Medical Ethics* 24: 263–268, 1998.

Galton, D. J. and Ferns, G. A. A. 'Genetic markers to predict polygenic disease: a new problem for social genetics'. *Quarterly Journal of Medicine* 92: 223–232, 1999.

Galton, D. J. and Galton, C. J. 'Francis Galton and his approach to polygenic disease'. *Journal of the Royal College of Physicians, London* 31: 570–574, 1997.

Galton, D. J. and Galton, C. J. 'Francis Galton and eugenics today'. *Journal of Medical Ethics* 24: 99–105, 1998.

Galton, D. J. and O'Donovan, K. 'Legislating for the new predictive genetics'. *Human Reproduction and Genetic Ethics* 6: 39–48, 2000.

Part 1: Towards the Genetic Modification of People

Introduction

Appleyard, B. *Brave New Worlds*. HarperCollins. London. 2000.

Bateson, P. and Martin, P. *Design for a Life*. Vintage. London. 1999.

Beveridge, W. B. *The Art of Scientific Investigation*. Random House. New York. 1950.

Castle, W. E. *Genetics and Eugenics*. Cambridge University Press. Cambridge. 1916.

Darwin, M. L. *What is Eugenics?* Watts & Co. London. 1928.

Duster, T. *Backdoor to Eugenics*. Routledge. New York. 1990.

Heisenberg, W. *Across the Frontiers*. Harper & Row. New York. 1975.

Kevles, D. J. *In the Name of Eugenics*. Harvard University Press. Cambridge, Mass. 1995.

Kevles, K. J. and Hood, L. (eds.) *Scientific and Social Issues in the Human Genome Project*. Harvard University Press. Cambridge, Mass. 1992.

Kitcher, P. *The Lives to Come*. Penguin. London. 1996.

Lewontin, R. C. *The Doctrine of DNA*. Penguin. London. 1993.

Peters, T. *Playing God*. Routledge. New York. 1997.

Rose, S. *Lifelines*. Allen Lane. London. 1997.

Rose, S., Lewontin, R. C. and Kamin, L. J. *Not in our Genes*. Penguin. London. 1990.

Sherrington, C. *Man and his Nature*. Pelican. London. 1955.

Silver, L. M. *Remaking Eden*. Phoenix. London. 1998.

Weir, R. F., Lawrence, S. C. and Fales, E. (ed.) *Genes and Human Self Knowledge*. University of Iowa Press. Iowa City. 1994.

Wilson, E. O. *On Human Nature*. Penguin. London. 1978.

Young, J. Z. *Doubt and Certainty in Science*. Oxford University Press. Oxford. 1951.

Prologue and Chapter 1: At Plato's Academy

Adam, J. *The Republic of Plato*, vol. 1. Cambridge University Press. Cambridge. 1965.

Aristotle. *Politics*, translated by J. Warrington. J. M. Dent & Sons. London. 1964.

Aristotle. *The Athenian Constitution*, translated by J. Warrington. J. M. Dent & Sons. London. 1964.

Bowra, C. M. *The Greek Experience*. Weidenfeld & Nicholson. London. 1957.

Bowra, C. M. *Ancient Greek Literature*. Oxford University Press. Oxford. 1959.

Davis, N. and Kraay, C. M. *The Hellenistic Kingdoms – Portrait Coins and History*, plates 13–47. Thames and Hudson. London. 1973.

Farrington, B. *Greek Science*. Penguin. London. 1953.

Herodotus. *The Histories*, translated by H. Carter. Oxford University Press. Oxford. 1962.

Kitto, H. F. D. *The Greeks*. Pelican. London. 1951.

Plato. *The Republic*, translated by F. M. Cornford. Oxford University Press. Oxford. 1955.

Plato. *Parmenides, Theaitetos, Sophist and Statesman*, translated by J. Warrington, J. M. Dent & Sons. London. 1961.

Plato. *Laws*, translated by A. E. Taylor. J. M. Dent & Sons. London. 1969.

Chapter 2: The Genetic Revolution: Naming the Parts

'A survey of the human genome'. *The Economist*, 1 July 2000.

Bodmer, W. and McKie, R. *The Book of Man*. Abacus. London. 1995.

Campbell, N. R. *Foundations of Science*. Dover Inc. New York. 1957.

Dawkins, R. *The Selfish Gene*. Oxford University Press. Oxford. 1989.

Dixon, P. *The Genetic Revolution*. Kingsway Publications. Eastbourne. 1997.

Jones, S. *The Language of the Genes*. Flamingo. London. 1994.

Jones, S. *In the Blood*. Flamingo. London. 1997.

Mueller, R. F. and Young, I. D. *Emory's Elements of Medical Genetics*. Churchill Livingstone. Edinburgh. 1995.

The Molecules of Life. Readings from the Scientific American. W. H. Freeman & Co. New York. 1985.

Wilson, E. B. *An Introduction to Scientific Research*. McGraw-Hill. New York. 1952.

Chapter 3: Designing Babies

Brownsword, R., Cornish, W. R. and Llewlyn, M. (eds.). *Law and Human Genetics*. Hart. Oxford. 1998.

Delhanty, J. D., Handyside, A. H. and Winston, R. M. 'Preimplantation diagnosis'. *Lancet* 343; 1569–1570. 1994.

Harris, J. 'Is gene therapy a form of eugenics?' *Bioethics* 7: 174–188, 1997.

Hindmarsh, R., Lawrence, G. and Norton, J. (eds.). *Altered Genes*. Allen & Unwin. St Leonards, NSW. 1998.

Jenner, E., Lister, J. and Pasteur, L. *Scientific Papers in Harvard Classics*. Collier & Son. New York. 1910.

Nicholls, E. K. *Human Gene Therapy*. Harvard University Press. Cambridge, Mass. 1988.

Pence, G. E. (ed.) *Flesh of my Flesh*. Rowman & Littlefield, Maryland. 1998.

Rifkin, J. *Biotech Century*. Phoenix. London. 1999.

Russell, S. J. 'Gene therapy'. *British Medical Journal* 315: 1289–1292, 1997

UNESCO. Press release no. 97–29, 28 February 1997.

Vogel, F. *Mutation in Man: Principles and Practice of Medical Genetics*, vol. 1, 2nd edn, edited by A. H. Emery and D. L. Rimoin. Churchill Livingstone. Edinburgh. 1990.

Weiner, N. *The Human Use of Human Beings*. Avon Books. New York. 1954.

Winston, R. *The IVF Revolution: The Definitive Guide to Assisted Reproductive Techniques*. Vermilion Press. London. 1999.

Chapter 4: Cloning Babies

European Parliament Resolution on Cloning, 11 March 1997 (passed 13 March 1997).

Galton, D. J., Kay, A. and Cavanna, J. S. 'Human cloning: safety is the issue'. *Nature Medicine* 4: 644, 1998.

Harris, J. *Clones, Genes and Immortality*. Oxford University Press. Oxford. 1998.

Josefson, D. 'US scientists plan human cloning clinic'. *British Medical Journal* 316: 167, 1998

Nussbaum, M. C. and Sunstein, C. R. (eds.). *Clones and Clones: Facts and Fantasies about Human Cloning*. Norton & Co. New York. 1999.

Public Perspectives on Human Cloning: A Social Research Study. The Wellcome Trust. London. 1999.

Shapiro, H. T. 'Ethical and policy issues of cloning'. *Science* 277: 195–196, 1997.

'US Senate bills on cloning under fire from researchers'. *Nature* 391: 623, 1998.

WHO document (WHA 50.37, 14 May 1997).

WHO press release (WHO/20, 11 March 1997).

Wilmut I., Campbell, K. and Tudge, C. *The Second Creation*. Headline. London. 2000.

Wilmut, I., *et al.* 'Viable offspring derived from fetal and adult mammalian cells'. *Nature* 385: 810–813, 1997.

Wise, J. 'Bills on human cloning are full of loopholes'. *British Medical Journal* 316: 573, 1998.

Chapter 5: Three Warnings from History

I: *Darwin and the 'struggle for life'*

Appleman, P. (ed.). *Darwin*. Norton & Co. New York. 1970.

Bodmer, W. F., and Cavalli-Sforza, L. L. *Genetics, Evolution and Man*. Freeman & Co. San Francisco. 1976.

Darwin, C. *The Origin of Species*. John Murray. London. 1901.

Darwin, C. *Autobiography*. Watts & Co. London. 1937.

Darwin, C. *The Descent of Man: And Selection in Relation to Sex*. Princeton Unversity Press. Princeton. 1981.

Darwin, F. (ed.). *The Autobiography of Charles Darwin and Selected Letters*. Dover. New York. 1958.

Dawkins, R. *The Blind Watchmaker*. Penguin. London. 1988.

Dobzhansky, T., Ayala, F. J., Stebbins, G. L. and Valentine, J. W. *Evolution*. Freeman & Co. San Francisco. 1977.

Fisher, R. A. *The Genetical Theory of Natural Selection*. Dover. New York. 1958.

Futuyma, D. *Evolutionary Biology*. Sinauen, Mass. 1986.

Huxley, J. *Evolution in Action*. Chatto & Windus. London. 1952.

Huxley, J. *New Bottles for New Wine*. Chatto & Windus. London. 1957.

Huxley, T. H. *Man's Place in Nature*. J. M. Dent & Co. London. 1906.

Jones, S. *Almost Like a Whale: The Origin of Species Updated*. Anchor. London. 2000.

Maynard-Smith, J. *Evolutionary Genetics*. Oxford University Press. Oxford. 1989.

Ritchie, D. G. *Darwinism and Politics*. Swan & Co. London. 1889.

Simpson, G. G. *The Meaning of Evolution*. Yale University Press. New Haven, Conn. 1949.

Stebbins, G. L. *Darwin to DNA*. Freeman & Co. San Francisco. 1982.

II: *Galton: Darwin's troublesome cousin*

Forrest, D. W. *Francis Galton: The Life and Work of a Victorian Genius*. Paul Elek. London. 1974.

Galton, D. J. 'Greek theories on eugenics'. *Journal of Medical Ethics* 24: 263–268, 1998.

Galton, D. J., and Galton, C. J. 'Francis Galton and his approach to polygenic disease'. *Journal of the Royal College of Physicians, London* 31: 570–574, 1997.

Galton, D. J. and Galton, C. J. 'Francis Galton and eugenics today'. *Journal of Medical Ethics* 24: 99–105, 1998.

Galton, F. *Natural Inheritance*. Macmillan & Co. London. 1889.

Galton, F. *Hereditary Genius*. Macmillan & Co. London. 1892.

Galton, F. *Inquiries into Human Faculty and its Development*. J. M. Dent & Sons. London. 1943.

King, L. (ed.). *A History of Medicine*. Penguin. London. 1971.

Leroux, L. 'A Vitrolles, une prime de naissance devenue torture morale'. *Le Monde*, 13 April 1998.

Pearson, K. *The Life, Letters, and Labours of Francis Galton*, vols. 1–3. Cambridge University Press. Cambridge. 1930.

III: *The Nazis and others*

Bormann, M. *Hitler's Table-talk, 1941–1944*. Oxford University Press. Oxford. 1988.

Bullock, A. *Hitler: A Study in Tyranny*. Penguin. London. 1962.

Burke, E. *Speeches and Letters on American Affairs*. J. M. Dent & Sons. London. 1961.

Burke, E. *The Philosophy of E. Burke*. University of Michigan Press. Ann Arbor. 1967.

Cohen, M. (ed.). *The Philosophy of John Stuart Mill*. Random House. New York. 1961.

Glasse, A. 'US and Norway used insane for Nazi-style tests'. *The Times*, 29 April 1998.

Haldane, J. B. S. *Heredity and Politics*. George Allen & Unwin Ltd. London. 1937.

Haldane, J. B. S. *Possible Worlds*. Harper & Row. New York. 1928.

Haldane, J. B. S. *Science of Life*. Pemberton. London. 1968.

Hilter, A. *Mein Kampf*, translated by J. Murphy. Hurst & Blackett Ltd. London. 1939.

Jefferson, T. *Democracy*, edited by S. K. Padover. Greenwood Press. New York. 1969.

Mill, J. S. *Three Essays*. Oxford University Press. Oxford. 1975.

'Nuremberg doctors' trial: 50 years on'. *British Medical Journal* 313: 1445–1475, 1996.

Toland, J. *Adolf Hitler*, vols. 1 and 2. Doubleday & Co. New York. 1976.

Chapter 6: The Individual and the New Eugenics

Burley, J. *The Genetic Revolution and Human Rights*. Oxford University Press. Oxford. 1999.

Cambon-Thomsen, A. 'Ethical considerations and French laws in bioethics: consequences for organisation of population genetic surveys in France'. Human Genome Meeting, Turin, Italy, 1998.

Confidentiality: Guidance from the General Medical Council, 1995.

Convention on Human Rights and Biomedicine. European Treaty Series 164. Editions du Conseil de l'Europe. Oviedo. 1997.

Knoppers, B. M., Hirtle, M., Lormean, S., Laberge, C. M. Caflanne, M. 'Control of DNA Samples and Information'. *Genomics* (forthcoming).

Rabinow, P. *French DNA: Trouble in Purgatory*. University of Chicago Press. Chicago. 1999.

Stern, K. and Walsh, P. (eds.). *Property Rights in the Human Body*. Kings College. London. 1997.

W. v Egdell (1990). All England Law Reports 835.

Chapter 7: The Family and the New Eugenics

Armstrong, C. 'Thousands of women sterilised in Sweden without consent'. *British Medical Journal* 315: 563, 1997.

'A role model of rigidity' . Editorial (concerning Mrs D. Blood), *Lancet* 348: 1253, 1996.

'Brides for seven brothers'. *The Economist*, 19 December 1999.

Convention on Human Rights and Biomedicine. European Treaty Series 164. Editions du Conseil de l'Europe. Oviedo. 1997.

Convention on the Rights of the Child. Editions du Conseil de l'Europe. Oviedo. 1989.

Galton, D. J. and Ferns, G. A. A. 'Genetic markers to predict polygenic disease: a new problem for social genetics'. *Quarterly Journal of Medicine* 92: 223–232, 1999.

Genetic Screening: ethical issues, 43. Nuffield Council on Bioethics, 1993.

Gottlieb, S. 'US couple files malpractice lawsuit against doctor for embryo mix-up'. *British Medical Journal* 318: 1025, 1999.

Harper, P. S. and Clarke, A. S. *Genetics, Society and Clinical Practice.* Bios Scientific. Oxford. 1997.

Human Genetics – Choice and Responsibility. British Medical Association, Oxford University Press. Oxford. 1998.

Lovestone, S., Wilcock, G., Rosser, M., Gayton, H. and Ragain, I. 'Apolipoprotein E genotyping in Alzheimer's disease'. *Lancet* 347: 1775–1776, 1996.

Medawar, P. B. 'The genetic improvement of man'. *Australian Annals of Medicine* 4: 317–320, 1969.

National Institute of Ageing/Alzheimer's Association Working Group. 'Apolipoprotein E genotyping in Alzheimer's disease'. *Lancet* 347: 1091–1095, 1996.

Nelkin, D. and Lindee, M. S. *The DNA Mystique.* Freeman & Co. San Francisco. 1995.

Silver, L. M. *Remaking Eden.* Phoenix. London. 1998.

Swinn, M., *et al.* 'Ethical dilemma: retrieving semen from a dead patient'. *British Medical Journal* 317: 1583–1585, 1998.

Tarasoff v Regents of the University of California (1976), 551 Pacific 2d 334 (Cal.).

'When a fetus is a person' (Dealing with Cornelia Whitner). *The Economist*, January 1998.

Chapter 8: Society and the New Eugenics

'A survey of the pharmaceutical industry'. *The Economist*, 21 February 1998.

Berger, A. 'Private company wins rights to Icelandic gene database'. *British Medical Journal* 318: 11, 1999.

Bindra, D. and Stewart, J. (eds.). *Motivation.* Penguin. London. 1971.

Brown, J. A. C. *The Social Psychology of Industry.* Pelican. London. 1970.

Goffman, E. *Interaction Ritual.* Penguin. London. 1967.

Low, L., King, S. and Wilkie, T. 'Genetic discrimination in life insurance: empirical evidence from a cross sectional survey of genetic support groups in the UK'. *British Medical Journal* 317: 1632–1635, 1998.

Lowie, R. H. *Social Organization*. Routledge & Kegan Paul. London. 1956.

McDougall, W. *An Introduction to Social Psychology*. Methuen. London. 1960.

Moore v Regents of the University of California 1990, 793 Pacific 2d 479.

Pokorski, R. J. 'Insurance underwriting in the genetic era'. *American Journal of Human Genetics* 60: 205–216, 1997.

Rose, A. M. (ed.). *Human Behaviour: A Social Process*. Routledge & Kegan Paul. London. 1962.

Russell, B. *Principles of Social Reconstruction*. Allen & Unwin. London. 1954.

Smith, P. B. (ed.). *Group Processes*. Penguin. London. 1970.

Worsley, P. (ed.). *Modern Sociology*. Penguin. London. 1972.

Chapter 9: Justice Without Imagination?

'A survey of human-rights law'. *The Economist*, 5 December 1998.

Brownsword, R., Cornish, W. R. and Llewlyn, M. (eds.). *Law and Human Genetics*. Hart. Oxford. 1998.

Charatan, F. 'Anti-abortionists ordered to pay $108m for threats of violence'. *British Medical Journal* 318: 415, 1999.

Commins, S. and Linscott, R. N. (eds.). *The Political Philosophers*. Random House. New York. 1947.

European Commission of Advisors on the Ethical Implications of Biotechnology. *The Ethical Aspects of Prenatal Diagnosis*. European Commission. Brussels. 1996.

'Freedom's journey: a survey of the 20th century'. *The Economist*, 11 September 1999.

Gaskin v UK (1989) Series A, 160.

House of Commons Science & Technology Committee. *Human Genetics: The Science and its Consequences*. HMSO. London. 1995.

Kennedy, I. and Grubb, A. *Medical Law*. Butterworth. London. 1994.

Life Insurance and Genetics. A Policy Statement of the Association of British Insurers. London. 1997.

Locke, J. *Two Treatises on Civil Government*. J. M. Dent. London. 1955.

Macready, N. 'US state rules that a viable fetus is a person'. *British Medical Journal* 315: 1488, 1997.

Maine, H. J. S. *Ancient Law.* J. M. Dent & Co. London. 1931.

Montesquieu. *The Spirit of the Laws.* Cambridge University Press. Cambridge. 1990.

Paine, T. *Rights of Man.* Pelican. London. 1979.

Powledge, T. 'Genetic screening as a political and social development'. In D. Bergsma (ed.) *Ethical, Social and Legal Dimensions of Screening for Human Genetic Disease.* Stratton International Medical Books. New York. 1974.

'Protecting abortionists'. *The Economist*, 23 January 1999.

Russell, B. *Power.* Unwin Books. London. 1971.

Chapter 10: From 1984 to 2084

Abercrombie, M. L. J. *The Anatomy of Judgement.* Pelican. London. 1974.

Ayer, A. J. *Language, Truth and Logic.* Gollancz. London. 1955.

Ayer, A. J. *The Problem of Knowledge.* Penguin. London. 1956.

Ayer, A. J. *The Foundations of Empirical Knowledge.* Macmillan. London. 1964.

Bertalanffy, L. *General System Theory.* Penguin. London. 1973.

Buber, M. *I and Thou.* Scribner's Sons. New York. 1970.

Chekhov, A. P. *Letters on Literary Topics*, edited by L. S. Friedland. G. Bless. London. 1924.

Chekhov, A. P. *The Letters of Anton Chekhov*, edited by A. Yarmolinsky. Jonathan Cape. London. 1973.

Chekhov, A. P. *Seven Stories*, edited by R. Hingley. Oxford University Press. Oxford. 1974.

Chekhov, A. P. *Lady with Lapdog and other Stories*, edited by D. Magarshak. Penguin. London. 1976.

Chekhov, A. P. *The Island: A Journey to Sakhalin.* Greenwood Press. Westport, Conn. 1967.

Cohen, M. R. *A Preface to Logic.* Routledge. London. 1946.

Cohen, M. R. and Nagel, E. *An Introduction to Logic and Scientific Method.* Routledge & Kegan Paul. London. 1961.

Dewey, J. *Democracy and Education.* The Free Press. New York. 1944.

Dewey, J. *Reconstruction in Philosophy.* Mentor. New York. 1955.

Dewey, J. *The Quest for Certainty.* Putnam & Sons. New York. 1960.

Dewey, J. *Selected Writings.* The Modern Library. New York. 1964.

Freud, S. *Civilisation and its Discontents.* Hogarth Press. London. 1951.

Freud, S. *Totem and Taboo*. Routledge & Kegan Paul. London. 1960.

Fromm, E. *Man for Himself: An Inquiry into the Psychology of Ethics*. Fawcett Press. Greenwich, Conn. 1947.

Fromm, E. *The Fear of Freedom*. Routledge & Kegan Paul. London. 1960.

Galton, D. J. 'Anton Chekhov's MD thesis: Sakhalin Island'. *Journal of the Royal College of Physicians, London* 23: 189–193, 1989.

Gibbs, C. A. (ed.). *Leadership*. Penguin. London. 1970.

Ginsberg, M. *On the Diversity of Morals*. Heinemann. London. 1956.

Hahn, B. *Chekhov and Tolstoy in 'Chekhov'*. Cambridge University Press. Cambridge. 1977.

Harris, J. *The Value of Life: An Introduction to Medical Ethics*. Routledge. London. 1991.

Harré, R. *An Introduction to the Logic of the Sciences*. Macmillan & Co. London. 1965.

Hume, D. *Theory of Politics*. Nelson & Sons. Edinburgh. 1951.

Jahoda, M. and Warren, N. (eds.). *Attitudes*. Penguin. London. 1970.

Koenig, R. 'Biologists mobilize against anti-genetics referendum'. *Science* 275: 607–8, 1997.

Koteliansky, S. S. and Tomlinson, P. *Life and Letters of Anton Chekhov*. Cassell & Co. Ltd. London. 1928.

Mazzini, J. *The Duties of Man*. J. M. Dent & Sons. London. 1955.

McGregor, A. 'Swiss approve use of genetic engineering'. *Lancet* 351: 44, 1998.

McLuhan, M. *Understanding Media*. Abacus. London. 1973.

Orwell, G. *Nineteen Eighty-four*. Penguin. London. 1989.

Polya, G. *How to Solve It*. Princeton University Press. Princeton. 1973.

Popper, K. R. *The Logic of Scientific Discovery*. Harper & Row. New York. 1959.

Pribram, K. H. (ed.). *Perception and Action*. Penguin. London. 1969.

Rayfield, D. *Chekhov: The Evolution of his Art*. P. Elek. London. 1975.

Russell, B. *History of Western Philosophy*. Allen & Unwin. London. 1947.

Russell, B. *Human Knowledge: Its Scope and Limits*. Allen & Unwin. London. 1948.

Russell, B. *Authority and the Individual*. Allen & Unwin. London. 1955.

Russell, B. *An Inquiry into Meaning and Truth*. Pelican. London. 1965.

Russell, B. *The Problems of Philosophy*. Oxford University Press. Oxford. 1974.

Shelley, M. *Frankenstein*. Penguin. London. 1992.

Simmons, E. J. *Chekhov: A Biography*. Chicago University Press. Chicago. 1962.

Spinoza, B. *Ethics*. J. M. Dent & Sons. London. 1955.

Stebbing, S. L. *A Modern Elementary Logic*. Methuen. London. 1961.

Tarski, A. *Introduction to Logic*. Oxford University Press. New York. 1965.

Weiner, N. *Cybernetics*. MIT Press. Cambridge, Mass. 1962.

Weyl, H. *Philosophy of Mathematics and Natural Science*. Atheneum. New York. 1963.

Whitehead, A. N. *Science and the Modern World*. Cambridge University Press. Cambridge. 1938.

Whitehead, A. N. *Adventure of Ideas*. Cambridge University Press. Cambridge. 1947.

Whitehead, A. N. *Essays in Science and Philosophy*. Rider & Co. London. 1948.

Wittgenstein, L. *The Blue and Brown Books*. Harper & Row. New York. 1958.

Wittgenstein, L. *Tractatus Logico-Philosophicus*. Routledge & Kegan Paul. London. 1961.

Part 2: Which Genetic Markers?

Chapter 11: The Genetic Components of Disease

Berg, K., Rettestol, N. and Refsum, S. (eds.). *From Phenotype to Gene in Common Disorders*. Munksgaard. Copenhagen. 1990.

Bernard, C. *Introduction to the Study of Experimental Medicine*. Dover. New York. 1957.

Christiansen, F. and Feldman, M.W. *Population Genetics*. Blackwell. London. 1988.

CIBA Foundation Symposium 130. *Molecular Approaches to Human Polygenic Disease*. John Wiley & Sons. New York. 1987.

Crow, J. F. *Basic Concepts in Population, Quantitative and Evolutionary Genetics*. W. H. Freeman & Co. New York. 1986.

Emory, A. E. H. and Rimoin, D. L. (eds.). *Principles and Practice of Medical Genetics*, vols. 1 and 2. Churchill Livingstone. Edinburgh. 1994.

Falconer, D. S. *An Introduction to Quantitative Genetics*. Longman. London. 1985.

'From genome to health'. *Wellcome News* 20, 1999.

Galton, D. J. and Assmann, G. (eds.). *DNA Polymorphisms as Disease Markers*, NATO ASI Series vol. 214. Plenum Press. New York. 1991.

Galton, D. J. *Molecular Genetics of Common Metabolic Disease*. Edward Arnold. London. 1985.

Hartl, D. L. and Clark A. G. *Principles of Population Genetics*. 2nd edn. Sinauer Associates Inc. Sunderland, Mass. 1989.

Hart, D. L., Freifelder, D. and Snyder, L. A. *Basic Genetics*. Jones & Bartlett, Boston, Mass. 1989.

Huxley, A. *Brave New World*. Flamingo. London. 1994.

Jorde, L. B., Carey, J. C. and White, R. L. *Medical Genetics*. Mosby. St Louis, Miss. 1997.

Minkoff, E. C. and Baker, P. J. *Biology Today: An Issues Approach*. Garland. New York. 2001.

Potts, D. M. and Potts, W. T. W. *Queen Victoria's Gene*. Alan Sutton. Stroud, Glos. 1995.

Randle, P. J., Bell, J. and Scott, J. (eds.). *Genetics and Human Nutrition*. John Libbey. London. 1990.

Ridley, M. *Genome*. Fourth Estate. London. 1999.

Roberts, J. A. F. and Pembrey, M. E. *An Introduction to Medical Genetics*. Oxford University Press. Oxford. 1985.

Sasazuki, T. (ed.). *New Approaches to Genetic Disease*. Academic Press. Palo Alto, CA. 1988.

Shorrocks, B. *The Genesis of Diversity*. Hodder & Stoughton. London. 1980.

Suzuki, D. J., Griffiths, A. J. F. and Lewontin, R. C. *An Introduction to Genetic Analysis*. W. H. Freeman & Co. San Francisco. 1981.

Szekely, M. *From DNA to Protein*. Macmillan. London. 1980.

Wagner, R. P., Judd, B. H., Sanders, B. G. and Richardson, R. H. *Introduction to Modern Genetics*. John Wiley & Sons. New York. 1985.

Weatherall, D. J. *The New Genetics in Clinical Practice*. Oxford University Press. Oxford. 1985.

'What we learn from twins'. *The Economist*, 3 January 1998.

Chapter 12: Frail New World

The material of this chapter has been a research topic for the author during the past thirty years, so original papers as well as informative books have been quoted.

Books

Bearn, A. G. (ed.). *Genetics of Coronary Heart Disease*. Nordalis Trykker. Oslo. 1992.

Durrington, P. N. *Hyperlipidaemia: Diagnosis and Management*. Wright. London. 1989.

Galton, D. J. and Krone, W. *Hyperlipidaemia in Practice*. Gower Medical. London. 1991.

Galton, D. J. and Thompson, G. R. (eds.). *Lipids and Cardiovascular Disease*. Churchill Livingstone. Edinburgh. 1990.

Goldbourt, U., Faire, U. and Berg, K. (eds.). *Genetic Factors in Coronary Heart Disease*. Kluwer. Dordrecht. 1994.

Jacotot, B., Mathe, D. and Fruchart, J.-C. *Atheroscelrosis X1*. Elsevier. Amsterdam. 1998.

Lusis, A. J. and Sparkes, R. S. (eds.). *Genetic Factors in Atherosclerosis*. Karger. Basel. 1990.

Research Papers

Betteridge, D. J., Durrington, P. N., Fairhurst, G. J., *et al.* 'Comparison of lipid-lowering effects of low-dose fluvastatin and conventional-dose gemfibrozil in patients with primary hypercholesterolemia'. *Am. J. Med.* 96(6A): 45S–54S, 1994.

Brandt, B., Vogt, U., Schlotter C. M., *et al.* 'Prognostic relevance of aberrations in the erbB oncogenes from breast, ovarian, oral and lung cancers: double-differential polymerase chain reaction (ddPCR) for clinical diagnosis'. *Gene.* 159(1): 35–42, 1995.

Brunzell, J. D. 'Familial lipoprotein lipase deficiency and other causes of the chylomicronaemia syndrome'. In *Metabolic Basis of Inherited Disease* 1989: Scriver, C. R.

Chamberlain, J. C., Thorn, J. A., Oka, K., Galton, D. J. and Stocks, J. 'DNA polymorphisms at the lipoprotein lipase gene locus: associations in normal and hypertriglyceridaemic subjects'. *Atherosclerosis* 79: 85–89, 1989.

Cho, Y., Gorina, S., Jeffrey, P. D. and Pavletich, N. P. 'Crystal structure of a p53 tumor suppressor – DNA complex: understanding tumori-genic mutations'. *Science* 265(5170): 346–355, 1994.

Dammerman, M. and Breslow, J. 'Genetic basis of lipoprotein disorders'. *Circulation* 91(2): 505–512, 1995.

Giardiello, F. M., Brensinger, J. D., Peterson, G. M., *et al.* The use and interpretation of commercial APC gene testing for familial adenomatous polyposis'. *New Engl. J. Med.* 336: 823–827, 1997.

Hayden, M., Kirk, H., Clark, C., *et al.* 'DNA polymorphisms in and around the Apo-A1-CIII genes and genetic hyperlipidemias'. *Am. J. Hum. Genet.* 40(5): 421–430, 1987.

Hitman, G. A., Tarn, A. C., Winter, R. M., *et al.* 'Type 1 (insulindependent) diabetes and a highly variable locus close to the insulin gene on chromosome 11'. *Diabetologia* 28(4): 218–222, 1985.

Hobbs, H. H., Brown, M. S. and Goldstein, J. L. 'Molecular genetics of the LDL receptor gene in familial hypercholesterolemia'. *Hum. Mutat.* 1(6): 445–66, 1992.

Josefson, D. 'FDA approves genetic test for women with breast cancer'. *British Medical Journal* 316: 168, 1998.

Mattu, R. K., Needham, E. W. A., Morgan, R., Rees, A., Hackshaw, A. K., Stocks, J., Elwood, P. C. and Galton, D. J. 'DNA variants at the LPL gene locus associate with angiographically defined severity of atherosclerosis and serum lipoprotein levels in a Welsh population'. *Arteriosclerosis and Thrombosis* 14: 1090–1097, 1994.

Ormiston, W. 'Hereditary breast cancer'. *Eur. J. Cancer Care (Engl.)* 5(1): 13–20, 1996.

Peacock, R. E., Hamsten, A., Nilsson Ehle, P. and Humphries, S. E. 'Associations between lipoprotein lipase gene polymorphisms and plasma correlations of lipids, lipoproteins and lipase activity in young myocardial infarction survivors and age matched healthy individuals from Sweden'. *Atherosclerosis* 85: 55–60, 1992.

Raffel, L., Hitman, G., Toyoda, H., Karam, J., Bell, G. and Rotter, J. 'The aggregation of the 5' insulin gene polymorphism in insulin dependent (type I) diabetes mellitus families'. *J. Med. Genet.* 29(7): 447–50, 1992.

Ravelingien, N., Pector, J. and Velu, T. 'Contribution of molecular oncology in the detection of colorectal carcinomas'. *Acta. Gastroenterol. Belg.* 58(3–4): 270–3, 1995.

Rees, A., Shoulders, C. C., Stocks, J., Galton, D. J. and Baralle, F. E. 'DNA polymorphism adjacent to human apoprotein A-1 gene: relation to hypertriglyceridaemia'. *Lancet* 1(8322): 444–6, 1983.

Stocks, J., Holdsworth, G. and Galton, D. J. 'Hypertriglyceridaemia associated with an abnormal triglyceride-rich lipoprotein carrying excess apolipoprotein C-III-2'. *Lancet* 2: 667–71, 1979.

Stocks, J., Thorn, J. A. and Galton, D. J. 'Lipoprotein lipase genotypes for a common premature termination codon mutation detected by PCR mediated site-directed mutagenesis and restriction enzyme analysis'. *J. Lipid Research* 33: 853–857, 1992.

Thomas, D. J. B., Stocks, J., Galton, D. J. and Besser, G. M. 'Hypertriglyceridaemia and diabetes mellitus: cause or effect?' *Diabetic Medicine* 5: 85–86, 1988.

Zhang, Q., Cavanna, J., Winkelmann, B. R., Shine, B., Gross, W., Marz, W. and Galton, D. J. 'Common genetic variants of lipoprotein lipase that relate to lipid transport in patients with premature coronary artery disease'. *Clinical Genetics* 48: 293–298, 1995.

Chapter 13: Personality Traits

Barondes, S. H. *Mood Genes*. Penguin. London. 1999.

Boring, E. G. 'Intelligence as the tests test it'. *New Republic* 34: 35–36, 1923.

Brunner, H. G., Nelen, M. R., van Zandroort, P., *et al.* 'X-linked borderline mental retardation with prominent behavioural disturbance: phenotype, genetic localisation and evidence for disturbed monoamine metabolism'. *Am. J. Hum. Genet.* 52: 1032–1039, 1993.

Cattell, R. B. *The Scientific Analysis of Personality*. Penguin. London. 1970.

Cook, M. *Interpersonal Perception*. Penguin. London. 1971.

Gathercole, C. E. *Assessment in Clinical Psychology*. Penguin. London. 1968.

Gould, S. J. *The Mismeasure of Man*. Penguin. London. 1997.

Herrnstein, R. and Murray, C. *The Bell Curve: Intelligence and Class Structure in American Life*. Free Press. New York. 1994.

Hu, S., Pattatucci, A. M., Patterson, C., *et al.* 'Linkage between sexual orientation and chromosome Xq28 in males but not females'. *Nature Genetics* 11: 248–256, 1995.

Hudson, L. (ed.). *The Ecology of Human Intelligence*. Penguin. London. 1970.

James, D. E. *Introduction to Psychology*. Panther. St Albans, Herts. 1970.

Lazarus, R. S. and Opton, E. M. (eds.). *Personality: Selected Readings.* Penguin. London. 1970.

Lowe, G. R. (ed.). *Personal Relationships in Psychological Disorders.* Penguin. London. 1969.

Maher, B. (ed.). *Abnormal Psychology.* Penguin. London. 1973.

Montague, C. T., Farooqi, I. S., Whitehead, J. P., *et al.* 'Congenital leptin deficiency is associated with severe early-onset obesity in humans'. *Nature* 387(6636): 903–908, 1997.

Skinner, D. F. *About Behaviourism.* Random House. New York. 1974.

Stafford-Clark, D. *Psychiatry Today.* Penguin. London. 1956.

Talland, G. A. *Disorders of Memory and Learning.* Penguin. London. 1971.

Warr, P. B. (ed.). *Thought and Personality.* Penguin. London. 1970.

Williams, M. *Brain Damage and the Mind.* Penguin. London. 1970.

Index